24.95

THE THOMAS EDISON BOOK OF EASY AND INCREDIBLE EXPERIMENTS

Thos A Edison

THE THOMAS EDISON BOOK OF EASY AND INCREDIBLE EXPERIMENTS

The Thomas Alva Edison Foundation

WILEY

John Wiley & Sons, Inc.
New York Chichester Brisbane Toronto Singapore

The Thomas Edison book of easy and incredible
experiments / The Thomas Alva Edison Foundation.
    p. cm. — (Wiley science editions)
  Bibliography: p.
  Includes index.
  Summary: A collection of science and engineering
projects and experiments covering such areas as
magnetism, electricity, electrochemistry, chemistry,
physics, energy and radioactivity.
  ISBN 0-471-62089-0. — ISBN 0-471-62090-4 (pbk.)
  1. Science—Experiments. 2. Engineering—Experiments.
3. Science—Experiments—Juvenile literature.
4. Engineering—Experiments—Juvenile literature.
[1. Science—Experiments. 2. Engineering—Experiments.
3. Experiments.] I. Thomas Alva Edison Foundation.
II. Series.
Q182.3.T478 1988
507'.8—dc19

                                  88-20669
                                      CIP
                                      AC

ISBN 0-471-62089-0
ISBN 0-471-62090-4 (pbk.)

Printed in the United States of America

  10 9

PREFACE

As chairman of the Thomas Alva Edison Foundation, I am delighted to introduce to young people this new edition of our popular book of experiments. It is a compilation of booklets specially written for the Foundation in recent years by scientific and technical experts.

Although the individual booklets of experiments will continue to be available through the Edison Electric Institute, it was felt that there was a need for a one-volume collection. This led in 1987 to the publication of *The Thomas Edison Book of Easy and Incredible Experiments*. Recent changes in the administrative affairs of the Foundation, together with the need for a new supply of books under the imprint of our new publisher, suggested that a new preface would be in order.

As one of its many activities in support of educational opportunities in science, technology, and engineering, the Thomas Alva Edison Foundation began quite some years ago to arrange for the publication of experiments in the spirit of that superlative experimenter, Edison, that could be performed by intellectually curious young people.

As a rule to be followed by the experts entrusted with each booklet's preparation, we asked that the experiments be significant, using simple and inexpensive materials whenever possible. We wanted to give the young a hands-on feeling for what science and technology are all about, and a taste of the excitement that goes with these pursuits.

Many scientists, looking back, remember what drew them to science and made them enthusiastic young experimenters. It may have been a science book that opened their eyes, or a favorite science teacher, or a toy compass, a magnifying glass, or an encounter with the stars at night. In Edison's case, it was an elementary physical science book that his mother gave him when he was ten years old. He set up a small chemical lab in the cellar, did chemistry experiments, got electricity from voltaic jars, and built a miniature telegraph system.

It was to stimulate young minds in this fashion that the Edison Foundation provided the series of experiment booklets. But even we were surprised how they caught on with teachers and students. Here was something that went beyond reading and memorizing facts, important as these always are. Through the experiments, students entered the workshop of Nature. They found the excitement of making simple devices and getting them to work, and applying scientific principles they had learned in class. In turn, experiments encouraged further book and classroom study.

The Foundation distributed the booklets in quantity to teachers, schools, and others at very near to cost. In some years, the distribution was close to half a million copies. Many a young person has found in this way a new hobby, an absorbing interest, a lifelong career.

Living as we do in an age of big science and very large research expenditures, some may think that simple experiments with inexpensive equipment, even strips of a tin can, couldn't possibly be significant. Not so! Some amazing things have been discovered at little cost.

Isaac Newton (1642–1727) darkened a room, let just a little beam of light enter from the window, and put a simple optician's prism in its path. Out of the white sunlight that entered the prism came all the colors of the rainbow! From this he developed his theory of light.

Sir Alexander Fleming (1881–1955) was about to throw away a bacterial culture plate that had been invaded by a bluish mold called *Penicillium*. David L. Kirk of Washington University has described what happened then: "What Fleming saw, others undoubtedly had seen before: a clear region around the *Penicillium* mold in which no bacteria were growing. What Fleming realized, no one ever had realized before: The *Penicillium* was producing something that killed bacteria. He named this substance penicillin."

Aristotle (384–322 BC) was a renowned philosopher in ancient times, and apparently a very good teacher for he was chosen to serve as young Alexander the Great's private tutor. His writings on logic, physics, philosophy, government, art, and the good life are still studied today. Such was the stature of this man and the institutions that built upon his thoughts.

But Aristotle was not always right. Bertrand Russell (1872–1970) writes in one of his essays that Aristotle passed along to posterity the dictum that women have fewer teeth than men. To this Lord Russell responds that he should have asked Mrs. Aristotle "to keep her mouth open while he counted." Unfortunately, Aristotle's error was repeated for centuries—because Aristotle had said it. There was no appeal to Nature, no experiment. To look into someone's mouth was unintellectual.

Aristotle also wrote that heavy objects fall faster than lighter objects. This sounds reasonable, on the face of it, so writers and lecturers repeated that for almost 2000 years. But then a rather headstrong young Italian, Galileo Galilei (1564–1642), who was studying motions, forces, and mechanical things, did something highly unusual. He performed a series of *experiments* to find out whether or not Aristotle was correct.

What followed was one of the simplest, cheapest, and most important experimental discoveries in history. He looked around for some strong wooden boards that he set at a gently inclined plane. Then he borrowed some smooth metal balls of various sizes and weights. These he rolled down the boards to a finish line.

After many trials, he determined beyond doubt that when balls of unequal weight are released together from a starting line, they finished together at the end of the course, if they are heavy enough to be little affected by air resistance. The principle is dramatically illustrated in today's classroom when a feather and a steel ball are released simultaneously in an evacuated glass cylinder. They hit bottom together.

So began the modern science of mechanics. Galileo's "easy and incredible experiment" with a few boards and metal balls eventually helped to land astronauts on the moon.

Thomas Alva Edison (1847–1931) was the most persistent experimenter and the most successful inventor the world has ever seen. He and Galileo would have got on well together.

He came to maturity at a time when the United States was entering a period of tremendous physical and economic expansion. Applied science and technology were in great demand. The telegraph spanned the continent before the railroads did, but they were not far behind. The post Civil War nation was entering the age of steam and steel. It was a time when energy, ideas, and hard work could bring tremendous results.

Edison had these qualities. Moreover, he had a unique capacity for concentration and observation. Nothing escaped his notice. Unlike most of us, he took the time to really look at and think about the smallest details of anything that caught his attention. He was self-taught and developed remarkable mechanical ability. He was by profession an inventor.

As he came to realize how vast the field was that must be winnowed to find practical solutions to the projects that caught his interest, he developed the idea of a research laboratory where specialists worked as a team, a veritable invention factory. This team-of-experts approach to industrial research was one of Edison's greatest inventions.

One of the secrets of his success, in addition to genius, was his tireless experimentation. He and his colleagues would try and fail, try and fail, and write it all down. When after what seemed a thousand experiments a worker complained that they were getting nowhere, Edison pointed out that they had made substantial progress. They now knew a thousand experiments that didn't work. Eventually they succeeded—the invention of the carbon filament electric lamp!

He enjoyed talking and joking with the Press and liked to help them out with good quotes. His phonograph was his most incredible invention—it made him world famous over night—but he told reporters with a twinkle that it was easy in a way, the only one of his experiments that worked the very first time. He added, "I was as surprised as everyone else."

Having observed by close attention that more electricity passes through a powdered carbon column when it is touched with gentle pressure, he invented a much improved microphone for Alexander Graham Bell's (1847–1922) telephone, making it for the first time a practical, successful invention. Edison told the Press that he just had to improve Bell's gadget. Being a little hard of hearing, he explained, he couldn't hear over the darn thing! This truly great man liked to joke with the Press, and the reporters loved him for it.

Edison was an inspiration to many youngsters in my generation (I was born in 1897, just 18 years after the invention of the incandescent lamp in 1879). When I was about 11 years old, I spent my pocket money for sal ammoniac and wire. Out in the barn, I rigged up a wet battery with the chemical, producing about 1 1/2 volts. I wired this to a switch that would close when anyone opened the barn door. I strung the wires from the switch to my bedroom to my master switch and to a buzzer, then back to the barn, the battery, the switch on the barn door. Now if any horse thieves opened the barn door at night after I closed my master switch, the buzzer would sound the alarm. This was my first electrical project, quite in the style, I thought, of young Tom Edison, although I never caught a horse thief.

Years later, in 1922 after earning my mechanical engineering degree at Cornell, I was employed as a cadet engineer by the Public Service Electric Company in New Jersey. When a new unit was dedicated at the Essex Generating Station in Newark, Mr. Edison came to the ceremonies. I was introduced to him and shook his hand. Few of us are alive today who actually saw and spoke to the great man.

When in the late 1920s Henry Ford moved the Menlo Park buildings to Greenfield Village, I was given the assignment of erecting a 132,000-volt steel tower at the old site and putting a light on it. This tower of light can still be seen from trains going between New York, Philadelphia and Washington.

The list of Edison's inventions and discoveries is a long one. His kinetoscope led to the motion picture industry. His puzzling over the darkening of electric lamps after prolonged use led to the discovery of the Edison effect, basic in the development of the electronics industries. During his lifetime, he was granted 1093 patents and filled some 3400 notebooks with his ideas and experiments. There is a chronology of his life at the end of this book.

World War II was not only a battle between the military forces of the nations involved. It was a contest between productive industrial capacities of the opposing sides. Even more significant were the phenomenal contributions of science and technology to the eventual victory of the Allied forces.

When the United States joined in the hostilities, I was on loan to the War Production Board, charged with the allocation of materials and production capacity for the manufacture of heavy electric power equipment. In that position, I saw what it means for a great nation to convert, almost over night, to war production.

In 1943, I was called by General Eisenhower to duty at the front, to direct the restoration of electric, gas, and water facilities as the Allies advanced from Algiers to Italy and from the beaches of Normandy to Berlin. I saw first-hand the destruction that energy can wreak in war and vowed to myself that, God willing, after the war, I would devote my career to the development of energy resources for peaceful and constructive purposes. This I have done to the best of my ability, directing the energy and electric power aspects of the Marshall Plan, and continuing these international endeavors to the present day throughout the world.

Almost immediately after the war, a group of leaders in industry and education decided among themselves that one of the pressing needs of the future, to keep our nation strong, was a special endeavor to increase the numbers of young men and women going into science, technology, and engineering. They chose Thomas Alva Edison as their symbol. Clearly, the world needs more Edisons.

Under the leadership of Dr. Charles Franklin Kettering, "Boss Ket," engineer, inventor, General Motors executive, the Thomas Alva Edison Foundation was established in 1946 "to keep alive and active for the benefit of present and future generations the inspiration and genius exemplified in the life, accomplishments, and ideals of Thomas Alva Edison and thereby to stimulate research and educational activities for the more effective advancement of human welfare."

These words from the papers of incorporation are, as one might expect, a bit self-conscious and on the formal side. But the purposes of the Foundation have been carried out over the years with warm and sincere concern for the well-being of young people, for their right to fully develop their potential, and to assist all the dedicated teachers who help them along.

The worldwide acceptance of the Foundation's endeavors comes, I think, from the high caliber of our board of trustees, today representing not only this nation but many nations around the world.

Immediately after the Foundation was established, Charles Kettering was named president by acclamation. He continued to serve until his sudden death in 1958, when the trustees expressed their desire that I assume the responsibility. I have served as chairman and chief executive officer for over 30 years.

During his tenure, Dr. Kettering was the guiding force in shaping the specific programs of the Foundation. We still follow his principles today. He was a very wise and human individual. We often had breakfast together in those days in a downtown Detroit hotel. I remember once, when I expressed impatience with the slow progress that was being made on an important community project, he said with that gentle smile, "Walker, no matter how slowly something is moving, it is still going faster than that which is standing still." He had an unusual capacity for talking to young people. It was a delight to hear.

Another eminent trustee over many years was David Sarnoff (1891–1971), who came to the United States when he was 18 and in time became chairman of the Radio Corporation of America. One evening when we were sitting together at a dinner, he described the famous occasion, in June, 1912, when during his hours of duty as a radio operator for the Marconi Wireless Telegraph Company, at the most powerful radio station in the world atop the John Wanamaker department store in Manhattan, he picked up the distress signals from the sinking *S. S. Titanic,* remaining at his key for 72 hours, receiving and passing on the news to the world.

There are so many more I could mention if there were space: William Francis Gibbs, designer of the *S. S. United States;* Dr. Lee A. DuBridge, President of the California Institute of Technology; Dr. Harlan H. Hatcher, President of the University of Michigan; Mary Pickford, whom Thomas Edison was the first to call "America's Sweetheart"; Kenneth D. Nichols, District Engineer on the nuclear Manhattan Project and second in command to General Leslie R. Groves; Thomas J. Watson, Jr., of IBM; and many more. The support of such trustees has enhanced each activity sponsored by the Foundation. Among these are the following:

The Edison Science Institutes and Science Education Conferences—Conferences for science teachers, where they can exchange views with prominent scientists and keep up with progress in their fields.

Edison Science and Engineering Youth Day—Activities held in many communities here and around the world on Edison's birthday. Local industries and schools cooperate to show science students around laboratories and research facilities, where they meet the scientists and technicians involved and learn more about career opportunities.

The International Edison Birthday Celebration Symposium—Held annually in one selected city around the world where outstanding science students from a nation, and their teachers, are honored. The event requires three days and includes many classroom presentations by experts and visits to laboratories in the area.

Edison Colloquiums—Small seminars with educators and representatives from other fields to discuss broad and pressing needs in education.

Publications, books, films, and tapes on scientific subjects, teaching methods, and laboratory demonstrations.

It was the Foundation, under the leadership of Thomas J. Watson, Jr., that brought about the election of Thomas Alva Edison to the Hall of Fame for Great Americans at New York University in 1960. Honored there, he also lives on in the hearts of scientists and engineers who, while young, were inspired by his magnificent example and "caught the flame" that has lighted their paths ever since.

WALKER L. CISLER
Chairman,
The Thomas Alva Edison Foundation

Young Tom at the age of 14. Already he had sold candy and food on a train, published a small newspaper (*The Weekly Herald*) on that train, and "lived" in the public library during the train's all-day layovers in Detroit. "I didn't read a few books," said Edison. "I read the library."

CONTENTS

Preface   v

PART I: SIMPLE EXPERIMENTS IN ELECTRICITY, ELECTROCHEMISTRY, AND
          BASIC CHEMISTRY   1

      Experiment 1: A Simple Electrical Circuit........................   1
      Experiment 2: How a Doorbell Circuit Works.....................   2
      Experiment 3: How a Two-Way Switch Works   2
      Experiment 4: Conductors and Insulators.........................   3
      Experiment 5: Controlling Current with a Pencil.................   4
      Experiment 6: What is an Electrolyte?   5
      Experiment 7: Electricity from a Lemon...........................   6
      Experiment 8: The First Electric Battery   7
      Experiment 9: Gases from Electrified Saltwater   9
      Experiment 10: Ink for Secret Messages 10
      Experiment 11: Carbon Dioxide—the Fire Killer................. 11
      Experiment 12: Candy Crystals from a Sugar Solution............ 13

PART II: SIMPLE EXPERIMENTS IN MAGNETISM AND ELECTRICITY 15

      Edison's Carbon Experiments 15
      Experiment 1: The Variable Conductivity of Carbon 16
      Experiment 2: The Carbon Transmitter Principle................. 16

      A Model Telephone Transmitter 17
      Experiment 3: Building a Carbon Transmitter.................... 18
      Experiment 4: Testing the Transmitter 18

      Magnetism Experiments—the Telegraph 19
      Experiment 5: Making a Magnet and Identifying the Poles 20
      Experiment 6: Making an Electromagnet 21
      Experiment 7: Magnetism and the Compass 21
      Experiment 8: An Electromagnet with Two Coils 22

Edison's Home-Lighting Circuit 23
Experiment 9: Series and Parallel Circuits 23
Experiment 10: The Fuse in Action 24

Faraday Experiments 25
Experiment 11: Building an Electrophorus 26
Experiment 12: Building an Electroscope 27
Experiment 13: Faraday's "Ice-pail" Experiment 27
Experiment 14: Does Ice Conduct Electricity? 29
Experiment 15: Electroplating Your House Key 30

PART III: SELECTED EXPERIMENTS — from Edison's Phonograph to his Motion-
picture Camera 33

Reproducing Sound the Edison Way. 33
Experiment 1: A Phonograph Pickup 34

The Telegraph Relay 35
Experiment 2: The Relay in Action 35

The Relay Coil in a Fun Project 36
Experiment 3: A Secret Drawer Lock 37

Edison's Ore Separator 39
Experiment 4: Rejecting Counterfeit Coins 39

The Movies—Thanks to Edison. 40
Experiment 5: The Spirit of 1776. 41
Experiment 6: A Pinhole Camera. 41

Who Turned on the Lights?. 42
Experiment 7: Edison's Electric Light. 43
Experiment 8: A Light-bulb Indicator 43

Replacing the Bulb with a Bell 44
Experiment 9: A Portable Burglar Alarm 45

Edison's Power-Generating System 46
Experiment 10: A Speedy Electric Motor 46

PART IV: USEFUL SCIENCE PROJECTS—Electric Pens To A Simple Radio 51

Writing with Sparks 51
Experiment 1: An Electric Pencil. 53

Coded Messages with a Buzzer 54
Experiment 2: A Code Set 56

Light from a "Frankenstein" Battery. 59
Experiment 3: A Simple Battery 61

A Radio that Plays for Free 63
Experiment 4: A Basic Radio 64

A Supersensitive Cigar-box Microphone 66
Experiment 5: A Cigar-box Microphone 67

PART V: ENERGY FOR THE FUTURE 71

Getting the Facts 72

Experiment 1: Your Home and the Forces of Nature   73
Experiment 2: A Basic Home-energy Audit   74

Conserving Energy   77
Experiment 3: Can You Use Electricity More Wisely?   78
Experiment 4: Building and Weatherizing a Model House   81

Solar Energy   86
Experiment 5: Converting to a Solar Garden   86

PART VI: ALTERNATIVE ENERGY SOURCES   93

Solar Energy   93
Experiment 1: A Model Solar Hot-water Heater   94
Experiment 2: Electricity Directly from Sunlight   95

Wind Energy   97
Experiment 3: Converting Wind Energy into Electricity   98

Ocean Thermal-energy Conversion 100
Experiment 4: The Idea Behind Ocean Thermal-energy Conversion . . 100

Tidal Energy 101

Energy from Trash 102
Experiment 5: Turning Trash into Usable Energy 102

Coal Conversion 104
Experiment 6: Getting Methane from Coal 104
Experiment 7: Converting Coal to Fuel Gas 105

Geothermal Energy 105
Experiment 8: A Model Geothermal Steam Engine 106

The Fuel Cell 108
Experiment 9: Making a "Fuel Cell" 108

PART VII: NUCLEAR EXPERIMENTS 111

Experiment 1: An Oil-drop Model of a Splitting Atom 113
Experiment 2: A Domino Model of a Chain Reaction 114
Experiment 3: Observing Radioactivity with an Electroscope 115
Experiment 4: Observing Radioactivity by Radiography 117
Experiment 5: Observing Radioactivity with a Cloud Chamber 118
Experiment 6: A Model Nuclear Power Plant Steam Turbine 120
Experiment 7: Demonstration of how Radioactivity can be Shielded . 121
Experiment 8: Build a Geiger Counter (a Class Project) 123

Recommended Publications 129

Chronology of Events in the Life of Thomas Alva Edison 137

Index 143

Edison demonstrated an improved model of his tinfoil phonograph before the National Academy of Science in Washington, D.C., and to President Rutherford B. Hayes at the White House. This picture was taken in April, 1878.

PART ONE

SIMPLE EXPERIMENTS IN ELECTRICITY, ELECTROCHEMISTRY, AND BASIC CHEMISTRY

EXPERIMENT 1: A Simple Electrical Circuit

THINGS NEEDED: Penlight bulb. Small socket. Flashlight battery or large 1½-volt dry cell. A few feet of insulated wire, any kind. Switch (we'll build this item). Tin can. Tape. Tin snips. Hammer. Saw. File or rasp. Can opener. Screws. Screwdriver.

We may as well begin right now by making the switch. While we're at it, let's make two. We'll need the second switch later on for another experiment anyway.

Get a discarded tin can and wash it out. Completely remove both the top and bottom. Then cut the can lengthwise and flatten it out. Be very careful of those sharp edges. From this sheet of metal cut 2 strips ½" by 3" and 2 pieces ½" by ½". Don't throw away the leftover metal yet.

Next, cut 2 blocks of wood about 2" by 3". With screws, assemble each switch as shown in the drawing. Make sure that the contacting sections of the metal have been

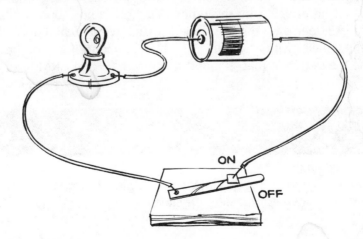

scraped clean. Also, leave enough room at the action end of the switch for another contact to be added. That will have to be done in a later experiment. Set one switch aside for the time being.

Now connect the circuit as shown. Use tape to hold the wires on the battery (soldering the wires in place would be better if you are able to do so). Note that when the switch is on, the circuit is closed and the bulb lights up. When the switch is off, the circuit is open and the current cannot flow.

EXPERIMENT 2: How a Doorbell Circuit Works

THINGS NEEDED: Same as in previous experiment, plus the second switch.

To set up the doorbell circuit, we'll have to convert both switches to push buttons. That's easy to do. Simply bend each arm slightly upward and position it over the contact. The light, of course, will serve as the bell.

With the circuit connected as shown, try each push button separately. Can you trace the completed circuit in each case? Note that when you "ring the back doorbell," it's as if the front-door switch wasn't even there. And vice versa.

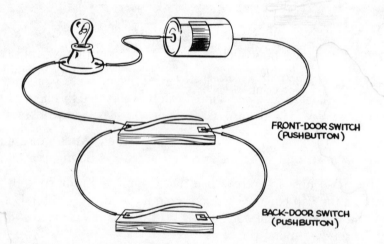

FRONT-DOOR SWITCH
(PUSHBUTTON)

BACK-DOOR SWITCH
(PUSHBUTTON)

These push-button switches are connected in *parallel* with one another. In a parallel circuit, there could be a dozen or more switches. Each one would work separately and not be affected by the other switches. Simple burglar-alarm systems work in this way. An open switch is positioned at each window and door. The switches are connected in parallel with an alarm. A slight movement of any of the windows or doors will close the switch at that spot and RRRRING THE ALARM.

EXPERIMENT 3: How a Two-Way Switch Works

THINGS NEEDED: Same as in Experiment No. 2, plus 2 more tin-can pieces ½" by ½". (Now you can throw the remaining tin-can scraps away.)

Have you ever been in a two-story house where the stair light can be turned on or off at either the top or bottom of the stairs? Some houses have a similar two-switch arrangement for the kitchen light when the kitchen has two doorways. No matter which way you enter or leave the room, you can control the light.

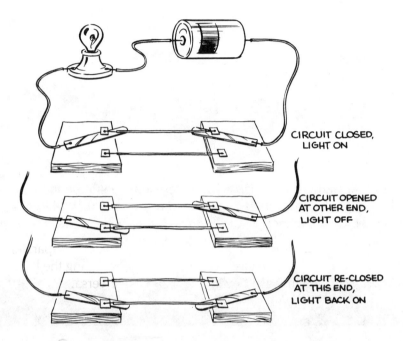

CIRCUIT CLOSED, LIGHT ON

CIRCUIT OPENED AT OTHER END, LIGHT OFF

CIRCUIT RE-CLOSED AT THIS END, LIGHT BACK ON

This experiment shows how such a circuit works. But first we must change our *single-throw* switches to *double-throw*. Do this by adding another contact as shown. Now each switch will make contact whether the arm is up or down.

Hook up the circuit according to the diagram and run a few tests. Note that when both switch arms are up, the circuit is closed and the bulb glows. The same is true when both arms are down.

But when one arm is up and the other is down, the circuit will be open. And the bulb will stay dark until one of the switch arms is moved. This explains why you can control the light from either switch no matter how the other switch is set.

EXPERIMENT 4: Conductors and Insulators

THINGS NEEDED: Same as in Experiment No. 1, except for the switch. Various test materials from around the house.

As we have found in the experiments so far, current will flow in a circuit as long as the circuit is closed. However, the materials in the circuit must be of the right kind. In other words, they must be able to conduct electricity.

All materials conduct electricity to some extent. The ones that pass electricity with ease are called "conductors." Silver, copper and aluminum are good conductors. Those that pass electricity with difficulty are called "insulators." Rubber, glass and porcelain are good insulators. Most materials, however, are neither good conductors nor good insulators but lie somewhere in between.

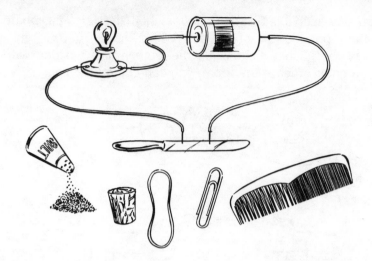

The circuit shown here will enable you to test some common items. Try to collect as many different kinds of materials as you can. Make sure that when you touch the material, you get good contact. If the bulb lights up, you know the material is a conductor. If the bulb doesn't light up, the material is not a conductor.

EXPERIMENT 5: Controlling Current with a Pencil

THINGS NEEDED: Same as in Experiment No. 1, except for the switch. Old pencil. Knife.

The "lead" in an ordinary pencil consists mostly of graphite, fine clay and wax. This mixture contains no lead. But because people began calling it lead, years ago, we still call it lead today.

A soft lead pencil has more graphite than a hard one does. Now although graphite itself can conduct electricity well, the mixture in a pencil conducts rather poorly. We're going to use a pencil lead to show that the length of a conductor affects the amount of current that flows in a circuit.

Carefully slit the pencil so that the lead is exposed. Now place both wires of the testing circuit near one end of the lead. The bulb should light up, although not nearly as brightly as with the wires touching one another.

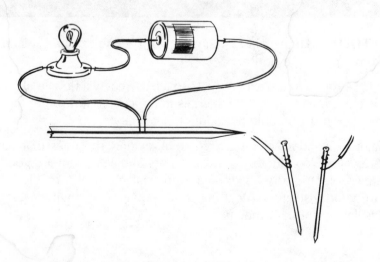

If nothing happens when you touch the lead, you'll have to press harder. One way to get extra pressure on the lead is as follows: Wrap each wire end tightly around the top of a long nail. Then use the nails as probes. In this way you'll be able to press quite hard. While pressing, slide one nail away from the other and notice what the bulb does. The farther apart the nails become, the dimmer the bulb gets.

What this experiment illustrates is that all conductors contain a certain amount of electrical resistance. As the conductor gets longer, this resistance can build up to a noticeable level. Even good conductors such as copper can't carry electricity too far without weakening the current.

This experiment also shows how a rheostat (REE-oh-stat) works. A rheostat regulates current flow. For example, the volume control on a radio is a rheostat. Room lights are often dimmed by a rheostat, somewhat in the manner that you dimmed the flashlight bulb…by running the contact arm over a longer length of conductor. This increases resistance, which reduces current flow.

EXPERIMENT 6: What is an Electrolyte?

THINGS NEEDED: Same as in Experiment No. 1, except for the switch. Various household liquids in separate containers.

That's an awesome word, "electrolyte" (eh-LEK-tro-lite). But don't let it scare you. It simply refers to a substance that is able to conduct electricity when dissolved in water. Dry salt cannot conduct electricity. But a solution of saltwater conducts it quite well. You might say, then, that an electrolyte is a liquid conductor.

For the test liquids, try separate water solutions of sugar, baking soda and starch. Also try water by itself. Then see what happens as you begin adding salt to the water. You might want to investigate the following liquids too: lemon juice, salad oil, vinegar and milk.

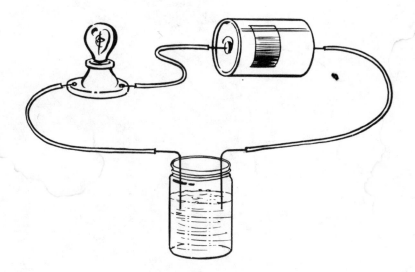

Why do some solutions conduct electricity while others do not? It has to do with the chemical compounds dissolved in the solutions. If the chemicals easily break up in the water into particles called "ions" (EYE-uns), the solution will conduct. Otherwise, the solution will conduct poorly or not at all.

But such theory belongs in high-school chemistry books. We mention it only because it is another example of the relationship between electricity and chemistry. Let us now try a few experiments involving both sciences. This field is known as electrochemistry.

THINGS NEEDED: A lemon. Small strip of copper about ½" by 3". Small strip of zinc, same size. (You can get zinc from the sides of an old dry cell, or you can use galvanized sheet metal—it's zinc-coated.) A galvanometer (this is easy to make; see the text). Scissors. Knife.

Here's an interesting demonstration in which a piece of fruit plays the role of an electrical cell. The act won't last long. And the amount of electricity produced is quite small. But it nevertheless is firsthand proof that chemistry can indeed cause current to flow in a closed circuit.

Since the current in this case is not nearly enough to light a bulb, we'll have to prove our results in some other way. We'll use a "galvanometer," which is a very sensitive instrument for detecting weak flows of electricity.

To make a galvanometer, we'll need a compass, a piece of stiff cardboard and a small spool of magnet wire (#28 or finer).

Cut a piece of cardboard as wide as the compass but long enough to fold up at the ends (see sketch). Then cut and fold another piece of cardboard the same as the first. Glue the 2 cardboard pieces back to back to form a cradle, as shown.

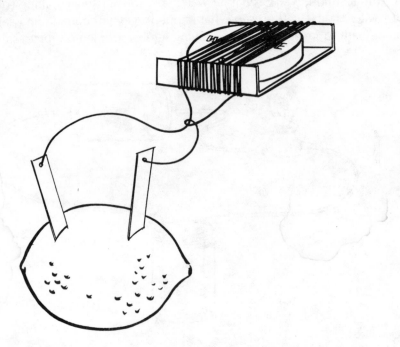

Place the compass inside the cradle so that the N-S axis points toward the closed sides. Finally, wind about 100 turns of wire around the compass and cradle right over the N-S axis. Twist the free ends a few times to keep the coil from unwinding.

Trim both ends of the coil to about 12" in length. Then scrape ½" of enamel insulation off the wire tips. And that's it. That's our galvanometer.

Here's how it works: Whenever current flows through the coils, it creates a magnetic field around the coils. With the coils aligned in the N-S direction, the field tends to swing the needle toward the E-S direction. A weak current will set up a weak field and swing the needle only slightly. A stronger current will swing the needle farther.

Now for the lemon. Roll it on a flat surface, pressing down with the palm of your hand. This will break up the inside of the lemon and let the juice flow freely. The lemon juice contains citric acid. This acid is the electrolyte for our lemon battery.

Next, cut 2 slits in the lemon about ½" apart and insert the copper and zinc strips, or electrodes, as they are called. Make sure that the electrodes do not touch each other inside the lemon. Finally, connect the galvanometer to the electrodes. The galvanometer must be level so that the compass needle can move freely. That should do it. The needle should indicate the flow of current, that is, the chemical production of electricity.

By the way, don't put the lemon in the garbage can yet. We will use it later when we make invisible ink.

EXPERIMENT 8: The First Electric Battery

THINGS NEEDED: Strip of copper about 1" by 6¼". Strip of zinc, same size. 2' of wire. 4 to 8 ounces of ammonium chloride.* Strip of thin felt ¾" by 6". Rubber band. Galvanometer from previous experiment. Hammer. Scissors. Steel wool or sandpaper. Bowl. Ruler. Tin snips.

While we're still on the subject of making electricity, let's build a battery. A battery is a group of 2 or more cells connected as one unit.

The battery we're going to make will be something like the one invented in 1880 by an Italian scientist named Alessandro Volta. His was the first electric battery...the first device for making electricity by chemical action. Today, simple chemical cells are called "voltaic" (vol-TAY-ik) cells in his honor. And when we use the word volt or talk of voltage, we also pay tribute to this man.

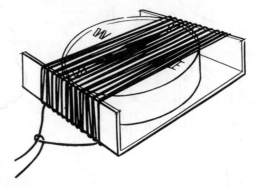

*Ammonium chloride is the same thing as sal ammoniac. That's what soldering blocks are made of. You can buy a sal ammoniac soldering block at hardware or department stores. It's not expensive.
Saw off about a third of it. Wrap this piece in a cloth, and break it up with a hammer till it's a powder. You can also get ammonium chloride powder from your druggist. But it's purer than we need and will, therefore, cost more.

Start by cleaning both sides of each strip of copper and zinc with steel wool or sandpaper. Then cut the copper strip into 5 squares 1" by 1". The piece left over, the sixth one, will be a little longer than the rest. Flatten the edges of all the pieces with a hammer. In the same way, cut the zinc strip and flatten the pieces. Also cut the felt into 6 pieces ¾" by ¾".

Dissolve the ammonium chloride into a few tablespoons of water in a bowl. This will be our electrolyte solution. We'll need enough solution to wet all 6 pieces of felt. And as soon as the solution is ready, you may as well put the felt pieces into the solution. They must be thoroughly soaked to begin with. But before they are used, you'll have to run your fingers down each piece, pinching the felt to remove most of the electrolyte. Otherwise, the completed battery will be wet all over. And that could short out some of the cells.

Before building the battery, punch a small hole in one end of the long copper piece and the long zinc piece. Attach a length of wire to the zinc by looping the wire through the hole and twisting it tightly (solder the connection if you can). Do the same with the copper strip.

Ready? First lay the long copper piece on a flat surface. On top of the copper, place one of the wet felt squares (don't forget to squeeze out the excess electrolyte). Next comes a zinc square. Then just keep repeating the cycle: copper, felt, zinc, copper, felt, zinc, etc. When you're done, the long zinc piece should be on top. At this point, secure the pile with the rubber band. You've just built a battery. Let's see how well it works.

Connect the battery to the galvanometer. The battery should cause a more lively response than did the lemon cell. In fact, the battery might even be strong enough to produce a faint glow in the penlight bulb. This battery will make electricity by chemical action as long as the felt squares remain wet with ammonium chloride.

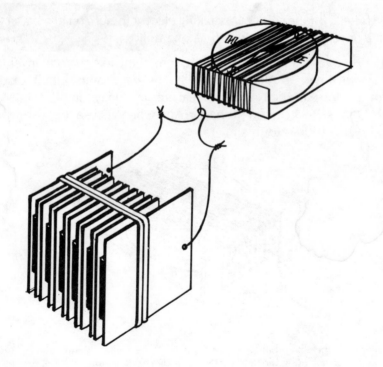

Here's another little experiment you can try if and when you take the battery apart. Keep the copper and zinc pieces from the ends of the pile. Leave the wires on.

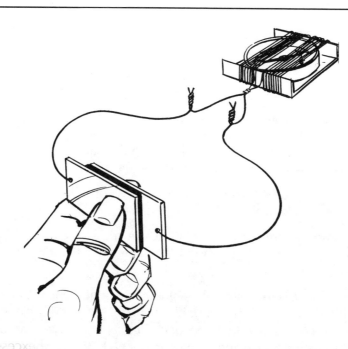

Clean the metal surfaces and put some fresh electrolyte on a new piece of felt. Then place the felt between the metal pieces. Arrange the pieces so that the wires go in opposite directions. Otherwise, the bare wire ends might touch and cause a short circuit.

After connecting the wires to the galvanometer, pinch the sandwich between your thumb and forefinger. You are holding a *voltaic cell*. Connect it to the galvanometer and see what it can do.

EXPERIMENT 9: Gases from Electrified Saltwater

THINGS NEEDED: Salt. 2′ of wire. Glass container or paper cup. 2 dead flashlight batteries. 2 fresh flashlight batteries. Saw or chisel. Tablespoon.

In the last two experiments we proved to ourselves that chemicals can be used to make electricity. Now we're going to prove that the reverse is also true...that

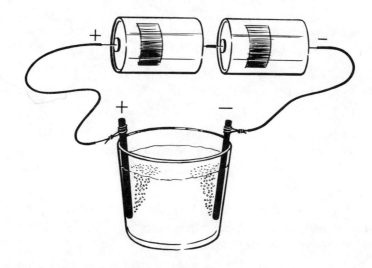

electricity can be used to produce chemicals. The action is something like forcing a cell to run backward.

Our cell in this experiment will have 2 carbon electrodes and a saltwater electrolyte. The forcing is going to be done by the flashlight batteries.

For the carbon electrodes we'll use the rods from the 2 dead flashlight batteries. Remove the rods from these batteries with a saw or chisel. Scrape the rods clean and wash them in clear water. Then connect the rods to the batteries as shown. Make sure that the wires are looped around the rods tightly and twisted to hold them in place.

The container must be nonmetallic. Either a glass or a paper cup will do. Fill the container with water and stir in 2 tablespoons of salt.

When you immerse the carbon electrodes in the saltwater electrolyte, do not allow the solution to touch the wire. If that happens, metal will chemically enter the solution and affect our results.

As soon as the electrodes touch the solution, current will start flowing between them. You will then notice many small bubbles streaming off both electrodes. On the electrode going to the negative (−) end of the battery, hydrogen forms. It is being unlocked from its union with oxygen in the compound H_2O, which of course, is water. On the electrode going to the positive (+) end of the battery, chlorine forms. It is being unlocked from its union with sodium in the compound known as sodium chloride, which is ordinary table salt. Electricity is causing this action to take place around each electrode.

Electricity also produces other chemical changes in saltwater. In industry, engineers use special equipment to take advantage of these changes. Along with hydrogen and chlorine, they are able to get sodium hydroxide from salt solutions.

Sodium hydroxide, also called caustic soda, is used for making soap and lye. The process of taking a solution apart, which is what we did in this experiment, is known as "electrolysis" (eh-lek-TROL-i-sis).

EXPERIMENT 10: Ink for Secret Messages

THINGS NEEDED: Lemon juice. Toothpick or ink pen. Candle. Sheet of plain white paper. Saucer.

Now that we've learned something about electricity and the way that electricity relates to chemistry, let's have some fun doing a few chemical experiments.

Still have that lemon from Experiment No. 7? We'd better use it before it gets rotten. Cut the lemon in half and squeeze out some of the juice into a saucer.

Using the lemon juice as secret ink and a toothpick as a pen, write something on the sheet of white paper. It might be a good idea to break the toothpick in half first. In that way you'll have a blunter writing tip. It will also carry a little more ink. Make sure to use enough ink to leave a wet trail. But write lightly with the toothpick or you'll dig a readable impression on the paper.

Notice that when the ink dries, it will be invisible. If what you have just written is a message to a friend, no one will be able to read it except your friend. How will he do it? He knows how to make the writing visible.

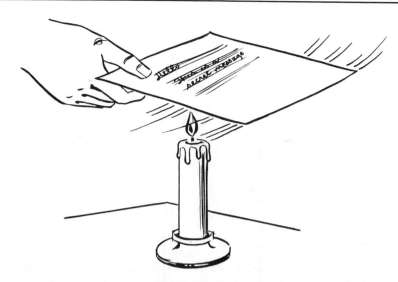

The trick is to heat the paper until the materials in the dried ink begin to scorch. This is done by passing the paper back and forth over a candle flame until the writing shows up. Care has to be taken not to hold the paper too long in one spot lest it ignite.

Why does heating the ink make it visible? Because of a chemical change. Just as heat chemically changes batter into cake, heat changes the ink into a new material. The ink, you see, burns at a lower temperature than the paper does. And so its brown ashes (the new material) stand out nicely against the white paper background.

You can also write invisible letters with grapefruit juice, milk and vinegar. Heat, again, makes the writing visible, for the same chemical reason.

EXPERIMENT 11: Carbon Dioxide—the Fire Killer

THINGS NEEDED: Baking soda. Vinegar. Large bottle such as a quart-sized pop bottle. Candle. Dishpan. Tablespoon. Small piece of paper. Measuring cup.

Carbon dioxide is a chemical compound containing 1 part carbon and 2 parts oxygen. Chemists write it as CO_2. It's one of our more common compounds. Human beings produce it, as a matter of fact, from the oxygen they breathe. Plants, in turn, absorb it to make oxygen. Soft-drink makers put it in pop to give it that gassy kick. And industry uses it to make dry ice, fire extinguishers and other products.

We're going to do two things in this experiment on carbon-dioxide gas. First, of course, we'll make it. Then we'll watch the gas smother fire...in this case, a candle flame.

Place the candle in a sink or dishpan. Be sure it is not near anything that could catch fire, like window curtains. Light the candle. Next, put about a tablespoon of baking soda into the bottle. An easy way to do this is to put the powder on a small piece of paper and then use the paper as a chute. Now pour a few ounces of vinegar into the bottle. Immediately the vinegar will start to foam. Place your thumb lightly over the bottle opening. You'll feel a pressure buildup. Why?

Baking soda is sodium bicarbonate. Vinegar contains acetic acid. When the two mix, a chemical reaction occurs, releasing carbon-dioxide gas. As the gas continues to form inside the bottle, the pressure on your thumb increases.

Before the foaming stops, remove your thumb from the bottle. Tilt the bottle to a horizontal position and place the opening slightly above the candle flame, being careful not to spill the liquid. (But don't worry if you do. It's harmless.)

If enough carbon dioxide has formed, it will slowly pour out of the bottle. You won't see it happen because carbon dioxide is colorless. Since it is about 1½ times heavier than air, the gas will begin to settle on the candle flame. When that happens, the gas will extinguish the flame. It does this for the simple reason that it deprives the flame of oxygen. And nothing burns without oxygen, as you know.

If you weren't successful in putting out the candle, start with a new supply of baking soda and vinegar. This time act a little faster and skip the pressure test with your thumb.

EXPERIMENT 12: Candy Crystals from a Sugar Solution

THINGS NEEDED: Sugar. Drinking glass. Cotton thread. Small pot. Pencil. Measuring Cup. Spoon for stirring.

What happens to grainy materials such as sugar and salt when they dissolve in water? The individual pieces change size. They go from grains to molecules...from something you can see with the unaided eye to something you couldn't see even with a powerful microscope.

A molecule of sugar, for example, is the smallest particle of sugar that can exist. If that particle were any smaller, it wouldn't be sugar. It would be some material whose molecules are not as big as sugar molecules.

In this chemistry experiment on solutions and crystals, we're going to convert sugar grains to sugar molecules. Then from the molecules we'll let the sugar re-form into large crystals—crystals big enough, clean enough and good enough to eat. In other words, we're going to make rock candy.

Put a cup of water into a small pot and bring it to a boil. When it starts to boil, turn off the heat, add about 1½ cups of sugar and stir. If all the sugar dissolves, add a little more. Keep adding sugar until no more will dissolve. The solution will then be *saturated*, meaning that it contains all the sugar it can hold at that temperature.

After the solution has cooled, pour it into a drinking glass. Next, hang a piece of cotton thread all the way into the solution. Support it from a string rod or pencil laid across the top of the glass. Then put the glass in a place where it will stay cool and undisturbed for a while.

As the solution cools, it becomes *super*saturated. Crystals will start forming on the string. After a few days, the crystals should be fairly large. You can remove the string and enjoy the sweet reward of a successful experiment.

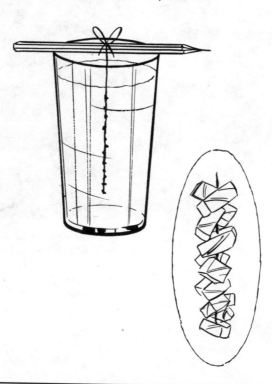

Edison with three of his "Edison effect" lamps. His unending experiments to improve the electric light resulted in his discovery, in 1883, that in an evacuated bulb, current could flow between two electrodes without wires.

The Edison effect lamp—which bears a striking resemblence to the pre-solid-state diode, or two-element vacuum tube—was the first invention to be patented that involved what we now call electronics. It thus became the cornerstone of that industry.

PART TWO

SIMPLE EXPERIMENTS IN MAGNETISM AND ELECTRICITY

EDISON'S CARBON EXPERIMENTS

Most people think of Thomas Edison as an inventor of mechanical and electrical devices. But he was more than an inventor. He was a discoverer as well. And many of his great contributions were not devices; they were observations...observations that blossomed because of his keen mind and his curiosity about the unusual.

One unusual thing that particularly interested Edison was the way in which pressure affects the electrical resistance of a mass of carbon particles. Like others before him, Edison noticed that when current flows through such a mass, applied pressure allows even more current to flow. And when the pressure is removed, the flow decreases. As we shall see, this observation later led to Edison's development of the telephone transmitter.

Edison first observed the variable resistance, or variable conductivity, of carbon in 1873. He pressed graphite, a form of carbon, in glass tubes and inserted wires in the graphite at both ends. And he found that the slightest pressure on the ends of the tubes changed the resistance.

Four years later he successfully exploited this curious property of carbon when he invented the carbon transmitter, or microphone. In his attempts to make transmitted voice sounds as clear as possible, Edison tried various forms of carbon, including graphite sticks and "carbon buttons" (made by pressing soot from kerosene lamps into little cakes). He even experimented with materials other than carbon.

Some of the experiments that follow will enable you to check carbon's conductivity properties for yourself. They may also give you some idea of how Edison carried out his persistent search for the solution to each problem he tackled.

For these experiments you will need a sensitive instrument to detect small current flows. As you already know from our previous experiments, such an instrument is called a galvanometer. It not only indicates the presence of current, but it also shows when the flow reverses direction. Turn to Page 6 to find out how to make your own galvanometer, if you haven't already done so.

EXPERIMENT 1: The Variable Conductivity of Carbon

THINGS NEEDED: Tube of powdered graphite. (Most hardware stores carry plastic squeeze tubes of graphite lubricant. Make sure it's graphite, not something else.) Small nonmetallic tube. 2 nails with wide heads (roofing nails). Paper napkin. 2' or 3' of hookup wire. 6-volt battery. Galvanometer (see Page 6). Tape.

First, cut 2 pieces of wire about 12" long and remove 1" of insulation from the ends. Take one of the wires and wrap a bared end around the shaft of one of the nails. Do the same with the other wire and nail. Try to make the connections tight, and use tape to keep them in place. Better yet, solder the connections if you can.

Next, wrap napkin strips around the nail shafts, enough so that each nail fits snugly (head first) into the nonmetallic tube. With a nail inserted in one end of the tube, fill the tube with graphite. Leave room for the second nail; then close the tube with that nail.

Now connect the tube wires to the galvanometer and battery, as shown. Note that when the circuit connections are completed, the current begins to flow, as indicated by the compass needle. If the needle doesn't move, check all connections. You may have to scrape some of the surface clean to make good contact. See what happens when you push the nails deeper into the graphite. Also, tap on one of the nails to make it vibrate and then observe the reading.

This experiment shows what Edison learned about carbon particles: A mass of such particles is electrically sensitive to movement and as a result, its conductivity is affected by pressure.

EXPERIMENT 2: The Carbon Transmitter Principle

THINGS NEEDED: Graphite, nail, wire, battery and galvanometer. Tablespoon. A penny. Tape. Flat stick.

Remove the tube from the circuit and take a nail off one of the wires (leave the other nail on). Tie the end of the free wire around the spoon handle; secure it with tape.

Prop something under the spoon handle to make the spoon level. Fill the spoon with graphite and gently smooth the graphite with a flat stick.

While watching the galvanometer, lightly touch the nail to the graphite. Touch several areas, varying the downward force. The compass needle should move each time, but not to the same degree.

Now place the penny on the graphite. Make sure that the coin does not contact the spoon. With the nail, touch the penny lightly. Note how much the galvanometer needle swings. Then press down on the penny with the nail. Again check the galvanometer. The needle should have swung farther, indicating an additional flow of current.

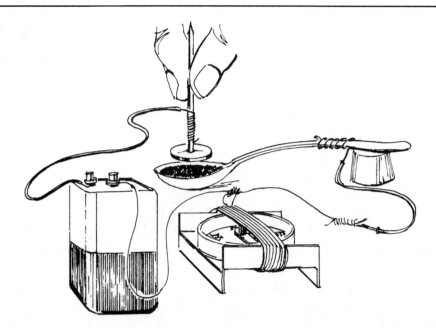

With this experiment you have demonstrated the basic principle used by Edison in his carbon transmitter. It is the same principle used in the mouthpiece of our telephones today. Pressure waves caused by your voice act on a diaphragm in the telephone (just as the pressure from the nail acted on the penny). The resulting vibrations vary the current passing through the carbon. These electrical variations correspond to the sound waves of your voice.

A MODEL TELEPHONE TRANSMITTER

Many of Edison's exploits consisted in devising something never before attempted or even thought of. However, he also made some important contributions by improving the inventions of other people, which helped make those inventions practical and widely useful. Take the telephone, for example.

Alexander Graham Bell was the first person to transmit speech over an electrical circuit. In Bell's original telephone, voice sounds directed into a speaker caused a diaphragm to vibrate. The vibrations of the diaphragm then induced weak electric impulses in a magnetic coil. The impulses traveled to the other end of the telephone line. There they produced similar vibrations in another diaphragm and were changed back to sound waves.

But the electric impulses were weak. Consequently, the speech carried only faintly and was difficult to hear and understand. Furthermore, the weak impulses limited Bell's range to only a few miles at best.

Edison helped make Bell's telephone a practical success by introducing the carbon transmitter. With his transmitter, Edison did not have to depend on the voice alone for the strength of the electric impulses. He used a carbon button to vary the flow of current from a storage battery. The current flowed from the battery to the carbon button in the transmitter, through an induction coil and then back to the battery. The induction coil boosted the voltage of the impulses produced by the voice vibrations.

These reinforced impulses were then put on the main line, where they could travel great distances and yet be heard clearly. Edison gave credit to his deafness for the improvements he made in the telephone. He said, "I had to improve the transmitter so I could hear it."

EXPERIMENT 3: Building a Carbon Transmitter

THINGS NEEDED: An all-metal frozen-juice can (or a can of similar size). Aluminum foil. Thin cardboard. Tape. 1' of hookup wire. Graphite from Experiment No. 1. (Carbon granules, such as those used in refrigerator deodorizers, would be even better.) Sandpaper. Rotary can opener. Scissors.

First we'll have to make a mouthpiece to talk into. The frozen-juice can will serve that purpose. Remove the top and bottom of the can with a rotary opener, to keep the edges smooth. Save one of the lids. Cut a circle of aluminum foil 1" larger than the can opening, and fasten the foil tightly over the can with tape. This will be our diaphragm.

Next, with sandpaper clean the entire surface of one side of the lid until the bare metal shows. Then sand a small spot on the other side of the lid. Solder the bared end of the 1'-hookup wire to that spot. If you don't have soldering equipment, use tape.

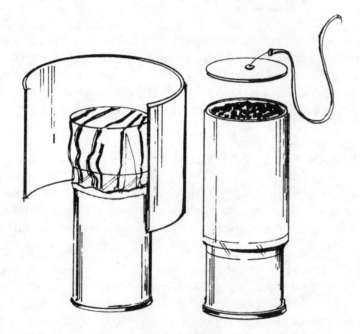

To hold the carbon, wrap a sheet of cardboard around the end of the can having the diaphragm. With about ½" or less of cardboard extending above the diaphragm, tape the cardboard to the can. Make sure that the overlapping edge of the cardboard is taped too, so the carbon can't spill out.

Fill the entire volume between the diaphragm and the top of the cardboard with carbon. (Better spread out some newspaper to avoid a mess.) To complete the transmitter, place the lid on the carbon so that the sanded side touches the carbon. Now tape the lid snugly in place against the carbon.

EXPERIMENT 4: Testing the Transmitter

THINGS NEEDED: Transmitter from Experiment No. 3. 1' of hookup wire. Galvanometer (see Page 6). 6-volt battery. Tape.

Connect the circuit as shown in the diagram. Before soldering (or taping) the wire to the side of the can, scrape that spot on the can down to the bare metal. Good connections are important. It is also important that the compass needle point to N at the start. When the connections are completed, the needle should move away from N and stay there. If it doesn't, check the connections and shake or tap the can a couple of times.

Once the needle moves, we can begin the test. Make any kind of loud noise into the open end of the can, and watch the needle.

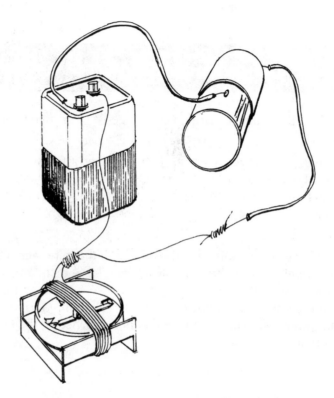

It should move back and forth. That's because the sound waves from your voice vibrate the diaphragm and cause a varying pressure on the carbon. And as you learned in Experiments 1 and 2, varying the pressure on carbon raises or lowers its electrical resistance, which changes the current flow. If you're not getting results, check the lid. Maybe it isn't pressing against the carbon tightly enough.

This experiment illustrates, once again, the principle of Edison's carbon transmitter, which made possible our long-distance telephones.

MAGNETISM EXPERIMENTS—THE TELEGRAPH

Originally the telegraph was a simple system for sending messages along a wire by means of electric impulses. For more than forty years in the mid-nineteenth century, it served as man's fastest means of long-distance communication. It was the reason for the end of the famous Pony Express.

When young Edison became interested in the telegraph, it could transmit only one message at a time on any one wire. And since the telegraph had been playing such a vital role in our country's economic growth, the system could not keep up with the flood of messages pouring in. Edison wanted to develop a system that could carry more than one message at a time.

He began to study all he could on the subject. He experimented endlessly. As a result, he not only made major improvements to the existing system, but invented the duplex, which handled two messages simultaneously; and soon after that, he invented the quadruplex, which handled four. Modern telegraph systems, of course, go far beyond this capability. But for that period (1874), four messages at once was quite an accomplishment.

Several experiments follow that illustrate some of the fundamental principles upon which Edison based his telegraph inventions. They involve the inseparable relationship between electricity and magnetism.

EXPERIMENT 5: Making a Magnet and Identifying the Poles

THINGS NEEDED: Bar magnet. Large sewing needle. Iron filings or crumbled steel wool.

Hold the sewing needle at its eye and stroke it repeatedly with the bar magnet toward the point. Always stroke in the same direction (not back and forth), and lift the magnet away from the needle for the return trip. After about 10 strokes, test the needle. You will find that it can pick up iron filings and other light pieces of metal. It is magnetized.

Tie a length of fine sewing thread to the middle of the bar magnet and suspend the magnet so it hangs freely. (Be sure to stand away from large metal objects.) The magnet will slowly align itself with the earth's magnetic field until one end is pointing to the north. That end is the north-seeking pole, generally called the north pole. Mark it with an N. Do the same with the magnetized needle, noting whether the north-seeking end is the eye or the point.

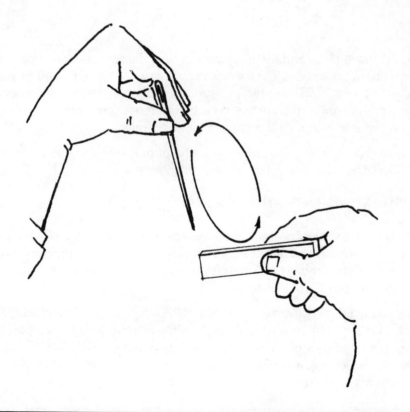

Now, with the magnetized needle suspended, advance the N pole of the bar magnet toward it. Notice the N end of the needle swing away from the N end of the magnet and the S end of the needle move toward the magnet. This action demonstrates the law of magnetic force: Opposite poles attract and like poles repel.

The reason a freely suspended magnet points its north-seeking pole toward the earth's arctic region is that the earth itself is a gigantic magnet, with its magnetic N pole near the geographic North Pole. Thanks to this happy circumstance, the compass exists.

EXPERIMENT 6: Making an Electromagnet

THINGS NEEDED: Iron bolt or spike. A few feet of hookup wire. Flashlight battery. Various small metal objects.

Wind about 20 turns of wire around the bolt, then connect the ends of this coil to the battery. The bolt is now an electromagnet and will attract metal objects as surely as your permanent bar magnet does. But when the battery is disconnected, the bolt will lose its magnetic force.

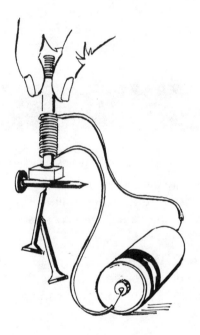

The strength of an electromagnet is determined in part by the number of turns of wire it has. The more turns of wire you give it, the stronger the electromagnet becomes, within limits. With the right amount of turns, our small electromagnet will probably be able to pick up a hammer. Electromagnets can be quite strong. For this reason, they are used in scrap yards to lift large piles of metal.

EXPERIMENT 7: Magnetism and the Compass

THINGS NEEDED: Electromagnet from Experiment No. 6. Compass.

Place the compass on a flat surface with the needle pointing to N. Connect the electromagnet and bring it near the compass. Note the direction in which the needle deflects.

Now, without moving the electromagnet, reverse the battery connections. The needle should deflect in the opposite direction.

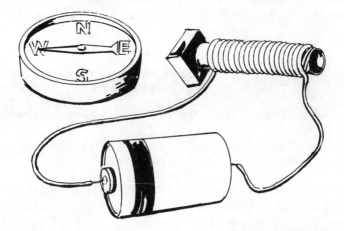

You have just repeated the historic discovery of a Danish schoolteacher named Hans Christian Oersted. He concluded that since current in a wire deflects a compass, it must be producing magnetism. It was one of the most important discoveries of all time. The motor operates on this law of nature.

EXPERIMENT 8: An Electromagnet with Two Coils

THINGS NEEDED: Same as in Experiment No. 7, plus more wire.

Wind another coil of wire around the bolt. Give it the same number of turns as before. Then connect the ends of the second coil to the same battery. With both coils energized, note what the compass does. Now reverse the connections on the second coil and again check the compass.

Here's what happened. When the current in both coils traveled in the same direction, the compass needle deflected as it did with only one coil. When the current in the coils traveled in opposite directions, the compass needle did not move at all, or

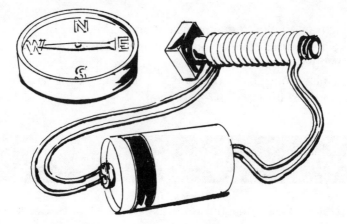

just slightly. That's because the magnetic fields set up by the coils were now aligned differently and the fields opposed, or balanced, one another. This experiment demonstrates that you can control the magnetism of the bolt by reversing the connections on either coil.

Through experiments like these, Edison found several methods of varying the magnetism in telegraph systems: change the direction of current flow through coils, change the amount of flow or change the direction in which the coils are wound. He combined these methods over the same wire. That was the duplex.

He then arranged to have two messages sent from the home station to a distant station while, at the same time and on the same wire, two messages were coming in from the distant station...four messages simultaneously. That was the quadruplex.

EDISON'S HOME-LIGHTING CIRCUIT

While Thomas Edison was trying to invent the incandescent electric light in 1879, he was also working on a system that would enable this light to be used in the home. That system included central power generators, switches, insulating materials, meters and many more items.

An important part of the system was a parallel wiring circuit for homes. Up to that time, the only electric lights in existence were the extremely brilliant carbon arc lamps that were beginning to be used for street lighting. These had to be connected in series. The following experiment shows the difference between the two types of circuits.

EXPERIMENT 9: Series and Parallel Circuits

THINGS NEEDED: 2' or 3' of hookup wire. 2 flashlight bulbs and sockets. Flashlight battery.

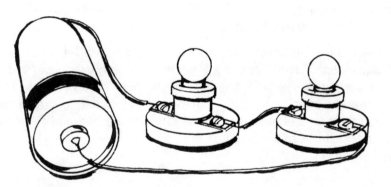

Start by setting up a series circuit, as shown. When the final connections have been made and the bulbs are lit, loosen either bulb. Note that no matter which one you loosen, when it goes out, so does the other one.

This is one reason Edison considered the series circuit impractical for his grand plan. Imagine having to light up an entire house just to read the newspaper, or having the entire house go dark because one bulb burned out.

Now change to a parallel circuit. Screwing either bulb out of its socket does nothing to the other bulb, which will continue to burn. Thus you can see, as Edison did, the advantage of a parallel circuit for home use. It allows individual lamps to be turned on or off without affecting other lamps in the circuit.

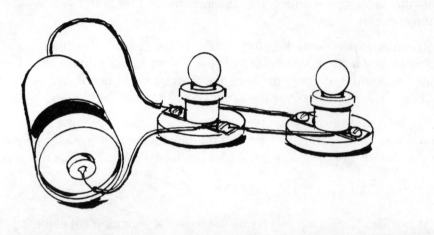

EXPERIMENT 10: The Fuse in Action

THINGS NEEDED: Wire, bulb, socket, and flashlight battery from Experiment No. 9 (you may need 2 batteries). A strip of Christmas-tree tinsel (silver icicle). Tape.

Undoubtedly the smallest, though not the least important, device in Edison's home-lighting system was the fuse. Something like an automatic safety switch, the fuse cuts off the current when it becomes high enough to cause a fire. Dangerously high currents in the main lines are the result of too many branch circuits in use at the same time (overloading the lines). Or they are the result of the "live" wire accidentally touching the "ground" wire or anything else that is grounded, such as a water pipe. This accidental touching is known as a "short circuit."

Edison's first fuse was patented on March 10, 1880, under the name of "Safety Conductor for Electric-Lights." He intended that such a fuse be placed in the circuit of each lamp or other electrical device.

It consisted of a piece of thin, special wire enclosed in a tube made of a nonconducting material. The wire had a low melting point. Whenever a short circuit (a high surge of current) developed, the heat of the high current would melt the wire immediately, thereby opening the circuit before any great damage occurred. The tube served to keep the droplets of molten metal safely contained and to prevent the two ends of the conductor from separating.

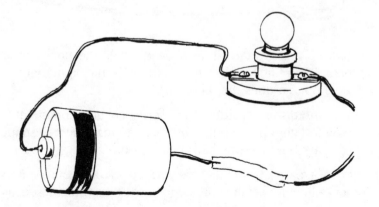

You can easily demonstrate how the fuse works. Lay 2 small lengths of wire parallel on a flat surface 1" or less apart. We're going to connect a piece of tinsel across the 1" gap. Use tape to make your connections, and be sure there is good contact at both ends of the tinsel. This will be our fuse.

Now comes the test. Connect the fuse wires with the lamp and flashlight battery, as shown. Use the tape to hold the wires to the battery ends. If everything is in order, the lamp will light.

To see the fuse in action, we'll have to produce a short circuit. Do this by touching the 2 terminals of the lamp socket at the same time with a pair of tweezers or another piece of wire. That will allow the current to bypass the lamp, taking a short cut, you might say.

IF A SECOND BATTERY IS NEEDED INSERT HERE

Without the lamp to act as a resistance, the current becomes much higher than it was. The load will probably be more than the tinsel can carry. If so, the tinsel will overheat, melt and open the circuit. However, you may have to use 2 batteries in series, depending on the thickness of the tinsel. If it weren't for our homemade fuse, the power source would spend itself in seconds.

We have this same kind of protection in our home fuses (and circuit breakers). Edison foresaw the possible danger in electrical overloads and short circuits. That's why he felt that the fuse was a necessary part of his system.

FARADAY EXPERIMENTS

More than anything else, Thomas Edison loved to experiment. He was unexcelled at it. Nikola Tesla, a European-born American inventor, wrote that if Edison wanted to find a needle in a haystack, he would examine straw after straw with the diligence of a foraging bee until he located the object of his search.

This intense interest in learning things for himself began at the age of nine, when his mother gave him a book on simple science experiments. Edison read every page and performed each experiment in the book with the methodical thoroughness for which he was to become renowned.

Years later he read two books on electrical experiments written by Michael Faraday, a great English scientist. Faraday made a lasting impression on him. In fact,

Edison regarded the purchase of Faraday's books as one of the decisive events of his life. "It was in Boston that I bought Faraday's works," Edison recalled. "I think I must have tried about everything in those books. His explanations were simple. He used no mathematics. He was the master experimenter."

The following experiments are based on research done by Faraday. Perhaps you will receive the same scientific inspiration in doing them as Edison no doubt did when he duplicated Faraday's experiments over a hundred years ago.

One of Faraday's most popular experiments is generally known as the "ice-pail" experiment. It represents the first proof that electric charges cannot be stored within a hollow conductor (Faraday used an ice pail) because they invariably travel to the outside surface.

To perform this experiment, you will first have to build an electrophorous (eh-lek-TROF-or-us) and an electroscope.

EXPERIMENT 11: Building an Electrophorus

THINGS NEEDED: A book. Sheet of polyethylene (from a dry-cleaning or vegetable bag). Piece of fur or wool. Bare metal disk (cover from a mayonnaise jar). Insulating rod (candle, glass, dry wood). Sandpaper.

The electrophorus is an electrostatic generator invented by Alessandro Volta in 1775. It consists of 2 parts: a charging plate and a charge-transfer disk. The simplest way of making the charging plate is to wrap a sheet of polyethylene around a hardcover book, keeping the top fairly smooth and tucking the edges under the bottom. You may want to build something more permanent, in which case your own resourcefulness is your best guide.

To make the charge-transfer disk, sand the top of the jar lid down to bare metal. Drop some melted wax from a candle onto the inside of the lid (if it has a waxed cardboard liner that is dry, leave it in; it will reduce the chance of charge leakage) and set the candle base on the wax before it hardens. If you don't have a candle, glue a glass rod or some other insulating handle to the lid. You now own an electrophorus. We'll discuss how it works in a moment.

EXPERIMENT 12: Building an Electroscope

THINGS NEEDED: Wide-mouthed jar or drinking glass. Thin cardboard to cover the jar opening. A 6" length of stiff, bare wire. Silver foil from a chewing-gum or chocolate-bar wrapper. Sheet of aluminum wrapping foil about 12" square. Tape. Large needle.

The electroscope is a sensitive instrument used for indicating the presence of an electrical charge. To build one, start by soaking the silver-foil wrapper in warm water to separate the foil from the paper liner.

While this is going on, cut the cardboard into a disk slightly wider than the jar opening. Use a large needle to pierce a hole in the center of the cardboard for the wire to pass through snugly. Next, squeeze the aluminum sheet into a ball. Then make an L-shaped hook in the wire about ½" from one end, insert the other end through the cover and jam it into the aluminum ball.

Now, back to the silver-foil wrapper. Peel the foil from the liner, dry it and smooth it out. Cut a strip 3" long by ¼" wide. Fold the strip in half and place it over the hook. Be sure that the ends hang side by side. If they don't, press them together gently.

Finally, lower the entire assembly into the jar and tape the cover in place. If the wire slides down into the jar, a few turns of tape around the wire will keep things in place. That takes care of the electroscope.

EXPERIMENT 13: Faraday's "Ice-pail" Experiment

THINGS NEEDED: Your electrophorus. Your electroscope. A metal container such as a large coffee can. Piece of fur or wool. Piece of flat glass (a framed picture, for instance). Strip of silver foil. Tape.

Before conducting this experiment, you should test the equipment you just built. Do it as follows: Charge the electrophorus by rubbing the charging plate vigorously with a piece of fur or wool. Then set the charge-transfer disk on the surface and touch the disk with your finger. This is an important step because your finger removes electrons and allows the disk to become positively charged. If you lift it up by holding only the candle, the disk will stay charged until you discharge it. Try bringing it close to your finger; you will draw a small, harmless spark and hear a slight snapping sound.

After recharging the disk, bring it near the foil knob atop the electroscope, but don't touch the knob. The foil leaves should spread apart. As you move the disk away from the knob, the leaves should slowly drop. The angle that the leaves make in the presence of a charge indicates the strength of that charge. The greater the angle, the stronger the charge. If the disk should happen to touch the knob, the leaves will remain apart after the disk is withdrawn. In that event, simply touching the knob will discharge the electroscope.

Now for Faraday's "ice-pail" experiment. As indicated earlier, Faraday devised this experiment to prove that electrical charges prefer the outside of a hollow conductor to the inside. You can prove it too.

Insulate the "ice-pail" container by setting it on glass (a framed picture will do fine). Hang a small strip of silver foil, about 3" by ½", on the outside of the can near the top. Attach it with tape. Now, with your electrophorus induce a charge in the charge-transfer disk (remember to touch it with your finger before removing it from the charging plate), and lower the charge disk into the can, without touching the sides. A charge will be induced on the inside of the can but it will transfer immediately to the outside, as indicated by the hanging foil, which will move outward slightly. You may have to look closely to perceive any movement of the foil.

Touch the disk to the inside of the can. This will discharge the disk. You may even see a spark jump. Make sure that the disk is discharged by bringing it close to the electroscope knob. The leaves should not react. Now, to verify that charges do indeed prefer the outside surface of a conductor, lay the uncharged disk against the outside of the can to pick up any charges (don't forget the finger trick). Then bring it near the electroscope again. The leaves should now diverge, proving the point.

For further proof, set the electroscope inside the can. Charge the outside of the can with the charged disk of your electrophorus. Although the foil strip on the can will move, indicating that a charge is present on the surface, the electroscope will show the interior to be neutral. Faraday's ice-pail experiment explains why it is safe to sit inside a closed car during a thunderstorm.

EXPERIMENT 14: Does Ice Conduct Electricity?

THINGS NEEDED: 2' of hookup wire. 2 strips of copper 1" by 3". Waxed paper or plastic container. 6-volt (or flashlight) battery. Galvanometer (to make your own, see Page 6).

Michael Faraday was deeply interested in electrical conductivity. And he measured the flow of electricity through many substances. One of these was ice. Concerning his experiments with this material, Faraday said, "I was working with ice when I was suddenly surprised to find that ice is a nonconductor of electricity."

This fact would probably surprise most people, since water is known to be a conductor (except when pure). But let's find out about ice for ourselves.

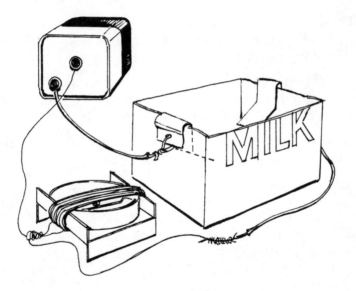

To the end of each copper strip, solder 1' of wire, bared at the ends. If you have no soldering equipment, punch a hole in the copper, loop the wire through and twist it tightly. Bend back the top ends of the copper so you can hook the strips on opposite sides of your container. Then angle the hanging part of the strips inward until the ends come within 1" or so of each other.

Form a series circuit by connecting one copper strip to the galvanometer, the other end of the galvanometer to either battery terminal, and the remaining battery terminal to the second copper strip. The only thing missing to complete the circuit is a conducting medium to electrically bridge the copper strips.

With ordinary drinking water, fill the container to about 1" from the top; at the same time, keep an eye on the galvanometer. You may or may not get a response, depending on what part of the country you live in. If the water in your area is hard, that is, if it has a high mineral content, it will pass current, as the galvanometer will show. But if the water in your area is soft, there's a chance that it won't conduct enough current in this particular experiment to deflect the compass needle (it can still conduct dangerous quantities in other circumstances though, so don't be misled).

If the needle on the galvanometer did move, showing that your local water conducts electricity, the next step is to place the container (with the copper strips still attached) in a freezer, or outdoors if the temperature is low enough. If your galvanometer needle did not move, you'll have to add salt to the water until it does. Then likewise freeze the container.

When the water has frozen solid, reinstall the container and the copper strips in the original circuit. You won't get a reading on the galvanometer because, as Faraday observed, ice is a very poor conductor. Not until a big enough path has been established through the partially melted ice will sufficient current flow to begin moving the compass needle. And as the path becomes bigger, the needle will increasingly deflect until it reaches a maximum.

EXPERIMENT 15: Electroplating Your House Key

THINGS NEEDED: Copper strip and container from Experiment 14. Salt and vinegar. 2' of hookup wire. Key or other metal object (plating will not affect its useability). Flashlight battery (a fresh one). Tablespoon.

In addition to his findings about electrical charges and conductivity, Michael Faraday made many other contributions to science. Certainly among the more important were his discoveries in the field of electroplating.

Electroplating is a process in which one metal is gradually deposited on another metal by means of electricity. Here is a simple experiment that shows the action involved.

Pour vinegar into the container until it is about half full. Add a tablespoon of salt, and stir. If all the salt dissolves, keep adding salt and stirring until the vinegar will dissolve no more salt (at which point the salt will begin settling to the bottom). This gives us the saturated solution we need.

Take 1' of the wire and solder one end to the copper strip and the other end to the button on top of the flashlight battery. If you don't have a soldering gun or iron, tape will also work. Similarly, connect the bottom of the battery to the key, or whatever else you decide to copperplate, using the other 1' of wire. Just looping the wire through the keyhole and twisting it will be sufficient. The key must be clean and dry.

Now immerse the copper strip and the key into the vinegar-salt solution, making sure that the two pieces of metal do not touch each other. Soon you will notice bubbles forming on the key and the color of the solution changing.

What is actually happening is that copper is being dissolved off the strip, driven across the solution by the current, and being deposited on the key. At the same time, hydrogen is being released from the water in the vinegar and is bubbling on the key. These bubbles must be wiped off periodically or the electroplating action will slow down. After a while you should be able to see a thin coating of copper form. Eventually you will have a nice, shiny copper-colored key.

Thomas Edison listening to his improved wax cylinder phonograph, after 72 hours of continuous work on the mechanism. This photograph was taken June 16, 1888.

PART THREE

SELECTED EXPERIMENTS—from Edison's Phonograph to his Motion-picture Camera

AN AGE OF GREAT INVENTION

Some of the experiments and projects in this section relate directly to several of Edison's most famous inventions. Others deal with some of the scientific principles that were integral in achieving the 1093 inventions he gave us. Before we begin our experiments, let's try to picture what life was like in Edison's time.

Back in the nineteenth century, even though the United States was still in its infant stages, it was regarded as the most inventive nation in the world. Samuel Morse had invented the telegraph, Alexander Graham Bell the telephone, Cyrus McCormick the reaper, and Elias Howe the sewing machine. Charles Goodyear had vulcanized rubber, George Eastman had patented photographic roll film, and even Abraham Lincoln had patented an invention of a novel riverboat.

But all through this revolutionary period of technical progress, the accomplishments of Thomas Alva Edison placed him head and shoulders above all other inventors.

Edison was a driving, self-taught American who used his intellect and burning curiosity to become the most productive inventor known to history. The people of his time regarded him as one of their greatest folk heroes. He was the man who was forever "making things."

Along with hundreds of other contributions, Edison gave us the phonograph, motion-picture camera, nickel-iron battery, electric railroad, multiplex telegraph, carbon microphone and, of course, the electric light. Without question, he belongs among the great builders of our country.

REPRODUCING SOUND THE EDISON WAY

From his early youth, Edison was hard of hearing. Perhaps that's why he devoted much of his time to the study of sound. He knew that the vibrations of the human eardrum, which is really a diaphragm, were communicated to the inner ear and that this caused the sensation of sound.

His experiments with megaphones, vibrating diaphragms and the telephone transmitter were stepping-stones to his invention of the phonograph in 1877.

As a boy, Edison, like many Americans, followed with intense excitement news of the pioneering bands of young telegraphers who extended their long lines of wire across the great prairies, through Indian-occupied territory and over the mountains to California. For, you see, the continent was first spanned not by the railroad, but by the telegraph, in 1861, during the opening months of the Civil War.

The telegraph revolutionized long-distance communication. Until its invention, messengers served as the chief means of communication over distance. Other means that had been tried—visual telegraphs, semaphores and signal fires from hill to hill— were poorly suited for widespread use.

The development of an efficient long-distance telegraph was extremely important to this nation's growth. While Samuel F. B. Morse in this country and Charles Wheatstone and William Coake in England are credited with the initial telegraph patents, the credit for vastly expanding the practical usefulness of the telegraph belongs to Edison. His many contributions include the duplex, quadruplex and multiplex telegraph systems, an automatic telegraph system, paraffin paper, the carbon rheostat and the telegraph relay.

Edison's phonograph reproduced sound as follows: A pickup stylus tracking the grooves of a recording (Edison used tinfoil wrapped around a cylinder) vibrated in accordance with the sound impressions in the grooves. Being attached to a diaphragm, the stylus caused the diaphragm to vibrate at the same rate. The diaphragm then emitted sound waves, which were made audible by a mechanical horn. Let's see if we can reproduce sound in the way that Edison did.

EXPERIMENT 1: A Phonograph Pickup

THINGS NEEDED: Frozen-juice can. Aluminum foil. Small cork. Sewing needle. Phonograph record (don't use a good one). Smooth cardboard. Ballpoint cartridge. Can opener. Rubber band. Scissors.

Horn and diaphragm. First remove the bottom of the juice can with a can opener. Then stretch a piece of aluminum foil tightly over both ends, folding the foil edges down the sides of the can and securing the foil with a strong rubber band. Try to get this diaphragm as taut as you can.

Stylus. To make the stylus, carefully tap the needle through the cork until the eye end is flat with the cork's surface. Glue the back of the cork to the center of the diaphragm. You have just built a mechanical sound pickup.

What's this I hear? With the can opening next to your ear, lightly scrape your finger against the stylus. Pretty loud, isn't it? Now, holding the pickup at an angle, let the needle rest lightly in the grooves of your record. Well I'll be a caveman's cousin. That's music coming out of the can.

Seeing sound. You can also demonstrate the vibrations that sound creates by converting your pickup to a recorder. Do this by replacing the needle with the ballpoint

cartridge. Then cut a round piece of smooth cardboard the size of your phonograph turntable. Punch a hole in the middle and place it on the turntable. Start the machine. As you hold the pen gently on the cardboard, make various sounds into the can. Note the wavy line caused by the vibrations of the diaphragm as the sounds from your voice strike it.

THE TELEGRAPH RELAY

You've heard of a relay, haven't you? It's an electromechanical device that allows a weak force to control a much stronger force. For example, a relay allows a switch in a safe, low-voltage circuit to open or close a higher-voltage circuit.

In the telegraph system that Edison operated and helped to develop, the current flowing in the line between two distant stations was too weak to operate the sounder. The sounder is what made the coded clicks that the operator would translate into a message. So to activate the sounder, a relay was used.

Here's how it worked: The telegraph line was connected to a coil in the relay that had many turns of wire around an iron core. When current flowed through the coil, meaning that a message was coming in, the coil became an electromagnet. The resulting magnetism caused two contacts in the relay to close. These contacts were part of a separate circuit that included the sounder and its own power supply. Variations in the intervals between clicks indicated what the message was.

Building and testing your own electromechanical relay will give you a clearer picture of this operation. In your setup, imagine the telegraph key (really a kind of on-off switch) to be miles away and the relay located at your end of the line. A light bulb will serve as the sounder.

EXPERIMENT 2: The Relay in Action

THINGS NEEDED: 2 wood blocks 4" by 2" by 3/4". A wood block 2" by 2" by 3/4". Some nails, including 2 roofing nails 1 1/2" long. 12' hookup wire. Metal strips. (It does matter what kind of metal strips are used. Neither copper nor aluminum would work. It must be a Ferris metal. An old "tin can" would work very well.) Popsicle stick. Thumbtack. 1 1/2-volt flashlight battery. 6-volt battery. 6-volt bulb. Bulb socket (the kind with wires already attached).

Relay frame. Using the drawing as a guide, nail the wood blocks together to form the frame. Then install a roofing nail about 1 1/2" from the open end. The top of the nail should be slightly below the top of the upright wood block. Also, mark a spot 1/4" from the open end for the other roofing nail, but don't pound that nail in yet.

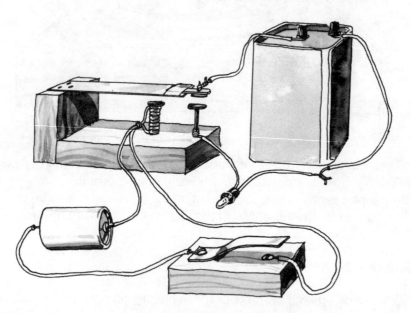

Electromagnet. Put about 100 turns of wire around the first roofing nail, leaving enough wire to make circuit connections. Keep the wire turns close together. When finished, twist the loose ends together so they won't come apart. Now you can drive the second roofing nail in.

Contact arm. For this you'll have to cut a 4" by 1" metal strip. Next, tape a piece of the popsicle stick to the end of the strip. Have the stick extend 1/2" beyond the strip. Near the end of the stick, press in the thumbtack, having the head on the same side as the strip. Make sure that the head is scraped clean. Then nail the strip to the frame so that the tack head is directly above the outermost roofing nail.

Telegraph key. Make this from the last wood block and a metal strip 4" by 1/2". Bend the strip as shown. Use roundhead screws for the terminals. Now you can connect all the parts to see the relay in action.

A message comes in. Note that when the low-voltage telegraph key is depressed (from miles away, remember?), the nail becomes an electromagnet and pulls the strip down. This closes the higher-voltage circuit, lighting the bulb. When the key is released, the electromagnetic field collapses, the strip springs upward and the bulb circuit opens.

THE RELAY COIL IN A FUN PROJECT

Edison used the relay principle in many of his inventions. It's easy to see why. This principle has thousands of applications.

Remember that we mentioned a coil in the relay that had many turns of wire around an iron core? We're going to use a coil similar to this in a fun project on the secret electromagnetic lock.

What is an electromagnetic lock? It's a vertical coil around a nonmagnetic tube (instead of the iron core in the telegraph relay) with an iron rod suspended in the center. The rod is supported by a piece of elastic thread and is arranged so that part of it extends above the tube. It serves the same purpose as a sliding bolt in an ordinary lock.

With the magnetic lock mounted on, say, the inside of a drawer near the top, the protruding rod prevents the drawer from being opened. But when a pair of secret terminals on the outside of the drawer are bridged, the coil yanks the rod down, thus allowing the drawer to be pulled out. When the circuit is broken, the rod will be lifted to its original position by the elastic thread. Power for the electromagnetic lock is supplied by a 6-volt battery kept inside the drawer.

EXPERIMENT 3: A Secret Drawer Lock

THINGS NEEDED: 1/4" nonmagnetic tubing 2 1/2" long. Copper or aluminum. 35' of hookup wire. An 8-penny nail. Elastic thread. Metal strip. 4 small nails. 2 machine screws with nuts. 6-volt battery. File or Rasp.

Coil. Wind about 300 turns of wire around the tubing. When you've finished, wrap some tape around the entire assembly so it won't come apart.

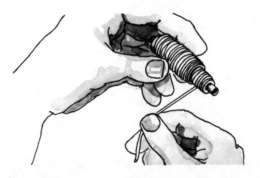

Rod and support. Make the rod by cutting a 2"-section from the middle of the 8-penny nail and file off any sharp edges. Then lay a 3" piece of elastic thread next to the rod, side by side. Tape them together at one end, as shown. Now lower the rod, taped end down, into the tube until only 1/2" sticks out the top. Place the remainder of the elastic over the edge of the coil and fasten it securely with tape.

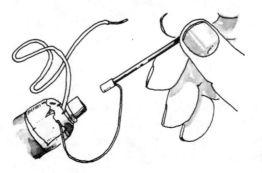

Installing the lock. Cut a metal strip 3″ by 1 1/2″ and form it around the coil. Bend the ends outward and punch two holes in each end to receive the nails. Before mounting the lock, energize the coil and measure the distance the rod drops. Suppose it drops 1/4″. This means that you should locate the lock on the drawer so that the rod extends 1/4″ into the space above the top of the drawer. If there is no space, you will have to make a small hole above the lock to receive the rod. When you have positioned the lock properly, nail it in place.

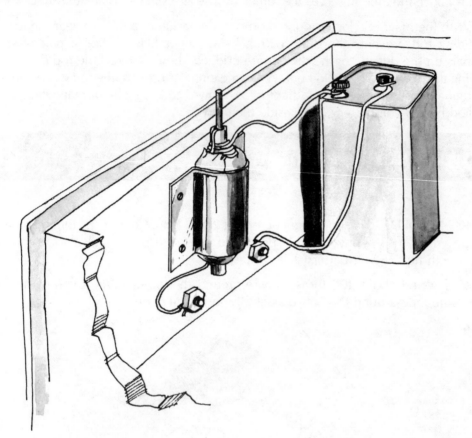

Connecting the circuit. The battery and lock are connected in series with 2 terminals (machine screws) that come through the drawer from the outside. Wires from the coil and battery are attached to these screw terminals with nuts.

Anything metallic that can touch both outside terminals at once will be your "electric key." It could be a coin, a scout knife, a metal-edged ruler. Of course, touching the terminals and putting on the drawer will have to be done at the same time. And it should be done quickly, to conserve the battery.

Free advice. Here's an idea on how to disguise the outside terminals. You could attach some sort of nonconducting (cardboard, wood, plastic) emblem or nameplate to the drawer front and use decorative screws as terminals. But even if you do nothing special with the screws, no one is likely to guess the secret of how the drawer is locked or what must be done to open it.

Incidentally, when the battery inside the drawer starts to weaken and can't quite pull the rod all the way down, you can still open the drawer by applying 6 volts (in the + to − direction) to the outside terminals, but don't let the battery die, for then you will have a little problem on your hands.

EDISON'S ORE SEPARATOR

While we're still on the subject of electromagnetism, let's take a look at one of Thomas Edison's inventions based directly on this principle. It's his magnetic ore-separation process.

Edison's reason for developing this process was to separate low-grade iron ore from the worthless material found with it. In the 1880s, a general belief was that iron ore was becoming scarce. Edison had access to 19,000 acres of land that contained, in his words, "over two hundred million tons of low-grade iron ore." He thought that it would be a simple matter to extract the iron and sell it at a good price. And that's what he set out to do.

It was no small operation. His iron mine and ore-processing plant at Ogdensburg, New Jersey, had miles of conveyors, gigantic rollers that could crush piano-sized rocks and the largest steam shovel in the country.

The heart of the whole system was a towering structure that housed the ore separator. In this building, powdered rock was hoisted to the top and allowed to fall past a series of screens and large magnets. The iron was deflected by the magnets into receiving bins; the sand dropped straight down to be carried away.

We could easily make a model of this separator by using sand and iron filings as the ore mixture. But we can demonstrate the identical principle in a less messy and more relevant way with a slug (counterfeit coin) rejector.

EXPERIMENT 4: Rejecting Counterfeit Coins

THINGS NEEDED: Iron bolt 2" long with nut and 2 washers. 20' of hookup wire. Thin, stiff cardboard 10" by 12". Telegraph key from Experiment No. 2. Various coins and coin-sized washers (slugs). 6-volt battery. Pencil.

Winding the electromagnet. Place a washer at each end of the bolt and engage the nut. Wind at least 4 layers of wire between the washers. After you've done this, twist the loose lengths of wire a few times to keep the electromagnet from unraveling.

Preparing the setup. First, put marks on the cardboard to indicate front, back, top and bottom. After that, draw a centerline down the length of both the front and back. Also draw a parallel line on the back 1/2" to the right of the centerline. Tape the electromagnet on this parallel line about 1/3 of the way from the top.

On the front, glue an A-shaped piece of wood as a divider. Locate it a few inches from the bottom and position the peak 1/4" from the centerline, toward the electromagnet.

Now we're going to convert the telegraph key to a switch. Very simple. Just cut a 3/4" metal disk, make a hole in the center, and put the disk under the free screw. Also straighten the arm and bend the end upward. The switch will be on when the arm is wedged under the disk.

Hey ma, it works! After connecting the circuit, prop the cardboard against something sturdy at about a 45-degree angle. From the top of the centerline, let several coins and slugs (washers) slide down the cardboard. If our slug rejector is working properly (bet you forgot to close the switch), all the slugs will be attracted to the electromagnet side of the divider and all the coins will drop straight down...somewhat like in Edison's ore separator.

That's because magnetism affects iron but not copper or nickel (or sand). Modern coin-vending machines detect slugs in a similar manner.

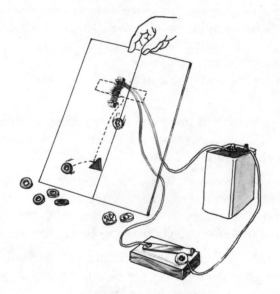

THE MOVIES—THANKS TO EDISON

During the middle years of his life, Thomas Edison's interest focused mainly on his secret "toy," the motion-picture camera. He could see no practical future for it, yet he could not bring himself to abandon the project. Perhaps he felt that it was a logical extension of his work on the phonograph, which he had invented years earlier and was still improving.

In 1888, Edison filed a document with the U.S. Patent Office stating: "I am experimenting upon an instrument which does for the eye what the phonograph does for the ear...this apparatus I call a Kinetoscope."

The Kinetoscope offered the first motion-picture show in history. With it, the viewer could look through a peephole and see a sequence of stop-and-go photographs that gave the illusion of continuous motion. Persistence of vision made the illusion possible.

Although today's motion-picture equipment is vastly superior to that which Edison developed, persistence of vision is still necessary for it to work. What is persistence of vision? It is the tendency of the eye to continue seeing an object for about 1/10 of a second after the object disappears. Here's a simple experiment to demonstrate the effect.

EXPERIMENT 5: The Spirit of 1776

THINGS NEEDED: 2" disk of cardboard. 12" of string. Glue.

Setting the stage. On the left half of the disk draw a large 17. Flip the disk over in bottom-to-top fashion and draw a large 76 on the right half. After laying a thin bead of glue across each side in a horizontal direction, stretch the string firmly and place it on the glue.

Seeing is believing. When the glue has dried, twirl the disk rapidly back and forth with the string. Thanks to the persistence of vision we all have, you will see our country's birth year as a complete number. In addition, you will see what lies directly behind the disk, which would normally be blocked out.

In developing his motion-picture camera, Edison used not only the persistence-of-vision effect, but certain optical principles as well. One of these explains how a camera sees an image, as shown in the next experiment.

EXPERIMENT 6: A Pinhole Camera

THINGS NEEDED: Large coffee can with plastic lid (the semiclear type). Large, dark cloth. Cardboard. Straight pin. Tape.

Basic camera. It's amazing how simple a camera can be, as you're about to find out. To make one, punch a hole about the size of a penny in the bottom of the coffee can. Do it in the center. Then cover the hole with a disk of thin cardboard, taping it down around the edges. Now pierce a hole in the disk with a pin, put the plastic lid on top of the can...and there's your camera, ready to go.

Screen and light shield. The plastic lid is our viewing screen. However, for you to see the projected image of what you aim at, the screen has to be shielded from light. For this purpose, drape the dark cloth over your head and the camera, the way some photographers do.

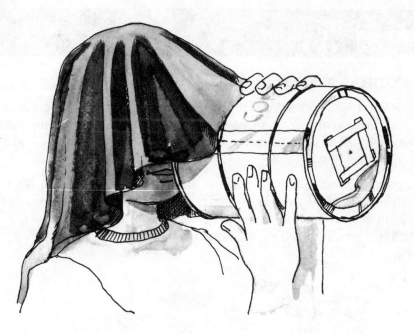

A topsy-turvy world. Aim the camera at something bright about 10" away. Hold the screen 1' or so from your eye. Note that what you see on the screen will be upside down. This is because light travels in a straight line. The light from the top of the object travels through the pinhole and lands on the bottom of the screen. Similarly, the light from the bottom of the object strikes the top of the screen.

WHO TURNED ON THE LIGHTS?

What do most people think of when they hear the name of Thomas Edison? The light bulb, of course. Invented by Edison in 1879, it was the most practical source of artificial light produced by man since the beginning of time. In the early 1800s, scientists in both the United States and Europe had experimented with hundreds of incandescent electric lights. But they couldn't get any of them to work, except perhaps for brief periods, until Edison succeeded.

As difficult as it was to develop, the incandescent electric light is basically a simple device. It consists of a coiled filament within a glass enclosure from which the air has been removed.

The principle on which it is based is that a glowing filament gives off both heat and light. The hotter the filament, the brighter the light. Edison's main problem was to find the right material to use for the filament. But he also had to create a good vacuum to prevent the filament from burning up.

After finding that carbon filaments worked best, he managed to make the filament last over longer periods of time by removing increasingly greater amounts of air from the bulb. We're going to try something similar in this experiment on the light bulb.

EXPERIMENT 7: Edison's Electric Light

THINGS NEEDED: Wide-mouthed jar with metal cover. 4' of hookup wire. Copper-strand lamp wire 4" long. Switch from Experiment No. 4. Birthday-cake candle on a small base. 6-volt battery.

Bulb. In the jar cover, punch 2 small holes just big enough to receive the wire. Space the holes 1 1/2" apart. Insert two 18" lengths of wire through the holes so they will extend halfway into the jar. Now bend the wires down the sides of the cover and tape them in place. Put a strip of tape over the holes too.

Filament. Remove 1 copper strand from the lamp wire. Wind it several times around a nail. Slip the coiled filament off the nail and connect it to the 2 wires coming from the cover.

Lighting up the dark. (Well...not quite.) Screw the cover on the jar. This is our "lamp." Next, connect the lamp in series with the switch and the battery. Turn the lamp on and start counting. The filament will begin to glow. If it continues glowing for more than 15 seconds, open the switch, otherwise you'll drain the battery. Try a shorter filament. Keep doing this until you find a length that burns for just a few seconds. When you do, put on a new filament of this length.

Now we're going to remove some of the air from the lamp. Put the candle inside the jar and ignite it. Then turn out the room lights. While the candle is burning, close the jar tightly. When the candle goes out, which means that it has used up a lot of the air, turn on the lamp again. Hopefully, the filament will glow a little longer this time. Letting it glow in the dark will produce a rather dramatic effect.

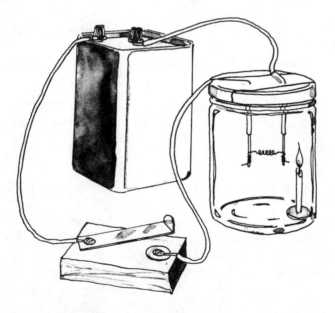

EXPERIMENT 8: A Light-bulb Indicator

THINGS NEEDED: 2 long, thin nails. 2' of hookup wire. Bulb socket from Experiment No. 2. Screw-type flashlight bulb. Flashlight battery. Tape.

We mentioned that the electric light bulb is a simple device. In addition, it operates in the simplest of circuits. An example is the light-bulb indicator you're about to put together.

Assembling the indicator. Actually, there's not much to assemble, as you can see. All you do is hook the light bulb and battery in series and attach the circuit ends to the nails. Make all connections by soldering (if you have no soldering equipment, use tape). Also cover the nails and nail connections with tape, leaving only the tips exposed.

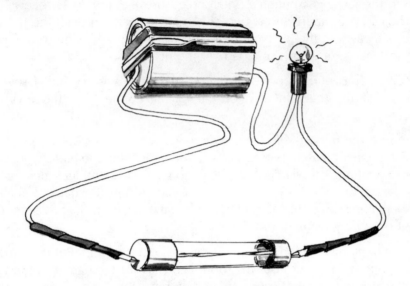

What do we do with it? Lots of things. For example, you can use it in science experiments to discover which materials and liquids are good (or bad) conductors of electricity.

You can also use it to check items such as flashlight bulbs or glass-tube fuses, as shown. Some of the fuses have wires so thin that you can hardly see them. If you touched the indicator nails to the ends of such a fuse and the bulb lit up, you'd know the fuse is okay.

But whatever you do, never use it on anything that is connected to a voltage source. (But then, you don't need to be told not to stick your finger in a beehive, do you?)

REPLACING THE BULB WITH A BELL

Similar to the light-bulb indicator and the homemade electric light is our next experiment, the portable burglar alarm. Can you see the similarity? Sure you can. All have the same three basic parts: a component that does something, a power supply, and a means of opening and closing the circuit. Also, all three are connected in a simple series circuit. The only difference is that the burglar alarm uses a bell instead of a bulb.

The burglar alarm provides a neat way of preventing a would-be intruder from opening an inward-swinging door without being heard. It is placed anywhere along the bottom of the door. When the door opens, it pushes against a contact arm. The arm, in turn, closes the circuit, which triggers the alarm.

EXPERIMENT 9: A Portable Burglar Alarm

THINGS NEEDED: 2 wood blocks 4″ by 3″ by 3/4″. 3 nails. Metal strip. 2 roundhead screws. 2′ of hookup wire. Doorbell. 6-volt battery. Hammer. Tape.

Frame and circuit contacts. After nailing the frame together, drive one of the screws partway into the vertical block about 1/2″ from the top. Then loop the bared end of a 6″ length of wire around the screw and finish driving the screw in.

Next, cut a strip of metal 5″ by 1/2″. After shaping the strip as illustrated, pierce a hole in the flat end. In a moment we will mount this contact arm at the bottom of the vertical block, using the remaining screw. But first let's assemble the alarm.

Tieing things together. Attach two 9″ lengths of wire to the doorbell terminals. Put the battery on the wood base, and hold the doorbell against the back of the battery. Now for a compact package, tie the bell, the battery and the vertical block together tightly with tape or cord.

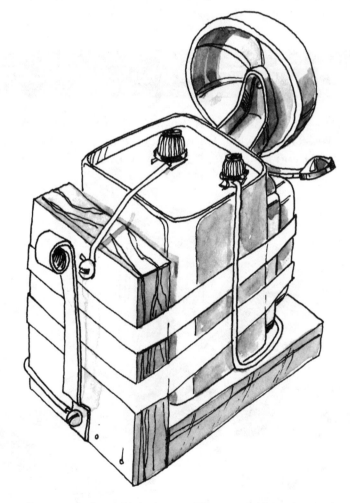

Final steps. Screw on the contact arm. Then complete these circuit connections: Run one doorbell wire to the battery, the other doorbell wire to the contact arm, and a third line from the upper contact screw to the battery.

"A Watchdog That Never Sleeps." The system should work like a charm. Anyone opening a door that is booby-trapped with this alarm will be in for quite a hair-raising surprise. Keep in mind, too, that this burglar alarm is truly portable. Since it has its own power supply, you can take it on family trips for use in hotels, motels or house trailers. Maybe everyone will sleep with more peace of mind.

EDISON'S POWER-GENERATING SYSTEM

As we said earlier, Thomas Edison knew that when he eventually reached his goal, the electric light wouldn't be of much use by itself. An entire central system for generating electricity and distributing it to homes would be needed. To Edison goes the credit for masterminding and building this system.

To make such a system a reality, Edison had to invent practically everything in it: underground distribution mains, insulating materials, power switches, meters, dynamos, lighting fixtures, fuses, and more. Certainly all of these inventions played an important role in the success of the system. But among the most vital was the world's first economical direct-current (DC) generator, or dynamo. This brings us to our last project in this section. The big one. It's a DC motor, which is essentially the same thing as a DC generator.

EXPERIMENT 10: A Speedy Electric Motor

THINGS NEEDED: Strap iron 6" by 5/8" by 3/32" (or 1/8"). 100' of #24 magnet wire. Two 16-penny spikes. Tape 1" wide. Darning needle or piece of coat hanger. Various metal strips. Copper-strand lamp wire 2" long. Base board 7" by 6" by 3/4". 4 screws. Some nails. 1' hookup wire. 6-volt battery. Vise.

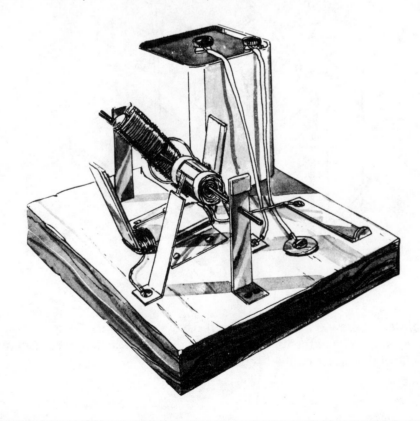

The motor we're going to build has 4 parts: the field magnet, armature, commutator and a pair of brushes.

Making the field magnet. We're going to need a vise for this job. Measure 2" in from each end of the iron strap. Then bend these 2" sections toward each other until the ends are about 3" apart.

Leaving 6" of wire for final connections, start winding the magnet wire along the 2" base of the iron strap. Put on 400 turns. When you've finished, wrap some tape around the windings to keep them in place and protect them.

Armature and shaft. The two 16-penny nails provide the backbone for the armature. Cut each of them to a 2 1/2" length, measuring from the top of the head down. Tape them as shown. As accurately as possible, find the midpoint of the assembly. Then insert the shaft (darning needle or 6" length from coat hanger) between the nails right through the tape, allowing 2" to stick out on the other side. Next comes the winding.

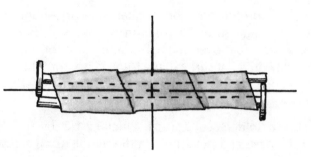

We will be putting on 4 layers of wire. Begin right next to the shaft. Leaving some wire for circuit connections, wind neatly toward the nail head. Keep each turn of the wire close to the preceding turn. Upon reaching the nail head, double back toward the shaft. Then make another trip to the same nail head, and return again to the shaft.

At this point, we cross over the other side of the armature without changing our coiling direction, and repeat the entire procedure. Two wire ends should now be extending from the center of the armature. Cut them down to about 2" and scrape the tips bare for connecting to the commutator terminals. (Note: Magnet wire has a deceptively clear varnish-like insulation on it, which must always be scraped off when making a connection.)

The commutator. For the terminals, cut two 1" by 1/2" metal strips. Shape them around any kind of 1/2"-diameter cylinder so that the length of the terminals runs with the length of the cylinder. Then, if you can, solder the armature tips to the ends of the terminals. Otherwise, pierce each terminal with a small nail, loop an armature tip through the hole and twist tightly.

Now we must build up the shaft diameter to hold the commutator terminals. Do this by wrapping 1" tape on the longer end of the shaft about 3/4" from the armature. Keep wrapping until a 1/2"-diameter cylinder is formed.

After looping the excess wire around the armature, place the commutator terminals on the rolled tape as shown. The space between the long edges of the terminals should be equal on both sides of the commutator. These spaces should face in the same direction as the nails. Put a thin strip of tape around each end of the commutator to keep it together. This completes the armature-commutator-shaft assembly. Okay so far?

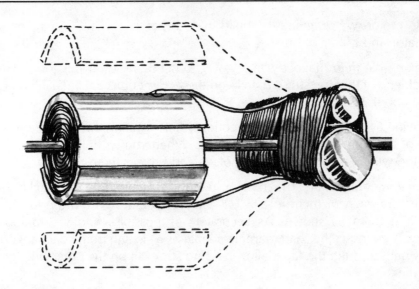

Supports for shaft and brushes. For the shaft supports, cut 2 strips of metal 3 1/2″ by 1 1/4″. Fold them in half so that the width is now 5/8″. Bend a 3/4″ segment 90 degrees for the base, and fold over about 1/4″ at the top for added stiffness. Standing each support vertically on a flat surface, measure 2″ up from the surface. Either drill or punch a hole at this point just big enough to receive the motor shaft. While you're at it, put a couple of holes in the base for the nails.

To make the brush holders, cut 2 strips of metal 2 1/2″ by 1/2″. In each one, bend a small segment for the base and put a hole in the middle for a screw. Two 1″ pieces of copper-strand wire will serve as the brushes. They should be soldered to the holders on the sides that will face the armature, with 1/2″ of copper extending above the holders. If soldering is not possible, try wrapping longer copper strands around the holders (scraped clean) and taping them on.

Putting the motor together. Mount the motor parts on the base board as shown on page 46. When doing so:

• Secure the field magnet to the board with a 2 1/4″ by 1″ metal strip and nails.

• Keep the clearance between armature ends and the field magnet fairly small.

• Use tape collars to keep the shaft from moving back and forth.

• Mount the brush holders with screws. Adjust them until the copper strands are vertical. Also, fan out the strands for better contact with the commutator.

Connecting the circuit. A good motor deserves a switch (and we didn't build a junky motor, right?). Make the switch arm from a metal strip 4″ by 1/2″. Cut a 3/4″ disk for the other contact. Use screws for the switch terminals.

In making the connections, follow the diagram (as you can see, it's a series circuit). Don't forget that the clear insulation must be scraped off the ends of the magnet wire.

Okay, start her up. Ready for the big moment? Let's keep our fingers crossed as we close the switch. You may have to give the armature a little push. If all the connections are good and nothing is binding the shaft, the motor should spring to life. Try adjusting the brush pressure for better performance. A little oil in the shaft support holes will help too.

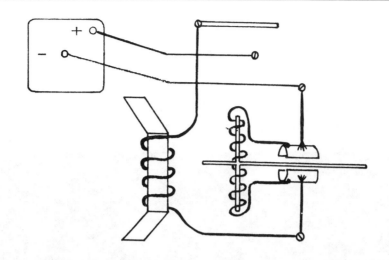

Just to see what our little workhorse can do, tie some thread around the shaft and let it hang over the edge of a table. See if the motor can lift, say, a lead pencil. Hey now, here's a possibility for your next science project in school.

STANDING ON THE SHOULDERS OF GIANTS

It is a commonly held concept that the inventions just "pop" into the mind of a Thomas Edison, that the inventor perceives a fully developed product in a flash of genius.

This "Eureka!-I-have-found-it" concept is not valid. At least it certainly was not the way that Edison produced his inventions. The incandescent lamp, together with the entire generating and distribution system needed to make his lamp workable, as well as his countless other inventions, were the results of long hours and tireless work.

Edison was fond of saying. "Genius is one-percent inspiration and ninety-nine-percent perspiration. Genius is hard work, stick-to-it-iveness and common sense. It is also preparedness."

Before starting any project, Thomas Edison would read everything he could find written on the subject. When asked, "To what do you owe your success?" he often replied, "I stood on the shoulders of giants," meaning, of course, that he had learned from the successes and failures of those who preceded him—men such as Michael Faraday, for example. The experiments in the next section incorporate the work by Faraday as well as that of Thomas Edison.

COPYRIGHT BY
THOMAS A. EDISON.

In the library of his West Orange, New Jersey, laboratory, Edison is using two of his inventions: the "Telescribe" (on the desk arm) and the "Ediphone" (on the stand). The Telescribe mates the Ediphone (a dictating phonograph) with the telephone so that when two people are talking on the telephone, the conversation of both parties can be recorded. This picture was taken on August 31, 1914.

PART FOUR

USEFUL SCIENCE PROJECTS—Electric Pens To A Simple Radio

WRITING WITH SPARKS

In 1875, Thomas Edison invented a device called the electric pen. He designed the pen for the purpose of writing words—in the form of fine holes—quickly and easily on special paper. The paper thus became a stencil.

Placing the stencil on a clean sheet of paper and then running an inked roller over the stencil produced a copy of the written words. The stencil could be used about 5000 times. This early idea of Edison's gave birth to a basic office machine, the mimeograph, that in modern form is still used today for reproducing letters and similar matter.

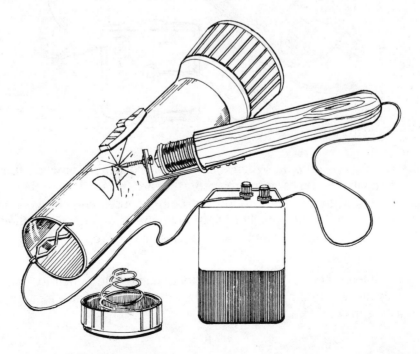

The electric pencil you are going to make is a cousin to Edison's pen. Instead of making holes in the paper, however, the electric pencil sparks a trail on metal. But being such a simple instrument, it will work only on soft metal.

HOW DOES THE ELECTRIC PENCIL WORK?

Our electric pencil operates on the same general principle as Edison's pen: the making and breaking of an electromagnetic circuit. As you already know, electricity flowing through a conductor causes an electromagnetic field to form around that conductor. For this priceless bit of technical knowledge, we can thank a Danish physicist, Hans Christian Oersted, who made the discovery back in 1819.

Bolt becomes an electromagnet. To see how the principle of electromagnetism applies to the electric pencil, let's look at the circuit diagram. Assume that the pencil tip has just touched the flashlight. This completes the circuit and allows current to flow (current is considered to flow from + to −). At that precise instant, a magnetic field builds up around the coil, making the core (the iron bolt) an electromagnet. The core, as a result, attracts the tip of the pencil and starts pulling it away from the flashlight. However, even though the circuit is being opened, the current tries to keep flowing. So it begins arcing from the tip.

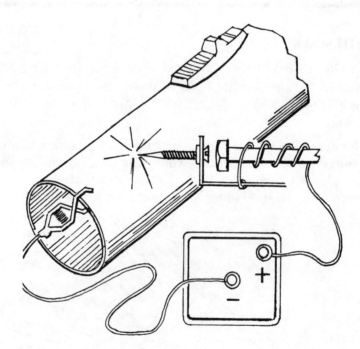

Broken arc stops current. When the increasing distance between the pencil tip and the flashlight becomes too great for the arc to continue, the arc breaks. Immediately the magnetic field around the coil collapses and the core releases its electromagnetic hold on the pencil tip. Tension on the flexed metal arm then snaps the tip back onto the flashlight. And the cycle begins all over, repeating itself many times each second.

It is the intense heat of the bluish arc that enables the pencil to "write." As the pencil moves along, the arc actually scorches the surface of the metal and thereby leaves a lasting trace.

EXPERIMENT 1: An Electric Pencil

THINGS NEEDED: HANDLE: Wood dowel 1" by 5" (the top of an old broomstick would be perfect). CORE: Iron bolt ¼" by 1¾". COIL: Spool of **#22** or **#24** solid hookup wire. ARM: Tin-can strip ½" by 3". TIP: Flat-head machine screw and nut ⅛" by ¾". BATTERY: 6 volts (12 volts would be even better, but not necessary). MINOR ITEMS: ¾" outer-diameter washer, 2 small screws, spring clamp, tape, file or rasp.

HOW TO MAKE THE PARTS

The handle and the core. Start by rounding off one end of the wood rod, if it's not already round. Drill the core hole in the other end as shown in the drawing. Make it slightly less than ¼" in diameter and about 1" deep. After slipping the washer around the iron bolt, twist the bolt into this hole. A little wax or soap on the threads will help. Leave 1" between the handle and the washers.

For a neat, professional-looking job, it would be nice to have a wire hole extending clear through the handle from the flat end to the round end. Use a ⅛" drill; bore in halfway from both ends. If you can't perform this operation, don't worry. You can run the wire along the top instead, holding it in place with tape or string.

Winding the coil. Insert the hookup wire into the hole (or run it along the top) until about 2' extend from the round end. Now begin winding the other end of the wire around the bolt. Put on as many layers as you can, finishing at the handle and trimming off all but 2" of the wire. We'll come back to the coil in a minute. But first the arm must be formed and installed.

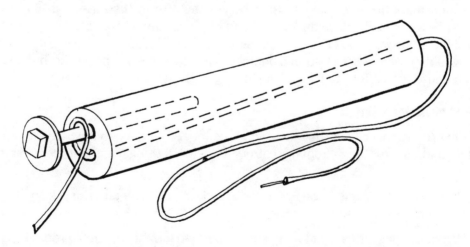

Forming the arm. Fold the ½" by 3" tin-can strip in half, lengthwise, so that it measures ¼" by 3". Doubling the tin-can thickness in this manner will give us the approximate flexing stiffness the arm needs. To fold the strip, simply scratch a line down the middle, clamp the strip in a vise at the line mark, and tap sideways with a hammer.

At about ¼" from one end of the folded strip, drill a hole big enough to receive the machine screw. Then, at the same end of the strip, make a 90-degree bend at a position such that when the long end of the arm is in place on the handle, the drilled hole will fall directly over the center of the bolt head.

ASSEMBLING THE PENCIL

First the machine screw must be filed to a point. Do this by placing the nut on the screw tightly, then letting the jaws of your vise grip the nut. When a long point is formed, remove the nut (this helps to reshape any roughened threads). Now insert the screw in the arm. Replace the nut and tighten.

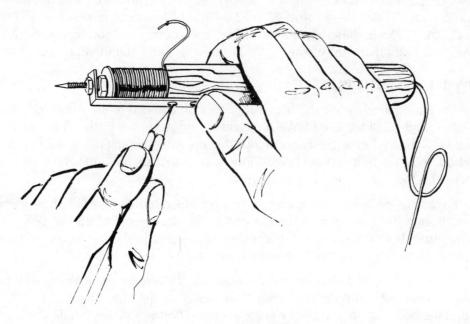

With your thumb, press the base of the arm against the bottom of the handle and position the arm so that about ¹⁄₃₂″ clearance exists between the bolt and the screw. At this point, mark the screw holes. Then secure the arm to the base with the 2 small screws. Leave enough space under the head of one of these screws to wrap the bared end of the 2″ tail from the coil. Now tighten it down. CAUTION: *Don't use long screws.* You wouldn't want a screw to accidentally dig into the wire inside the handle. This would cause the current to take a shortcut through the screw and bypass the coil.

THE PENCIL IN ACTION

To use your pencil, all you have to do is connect the pencil to the 6-volt (or 12-volt) battery and then run a line from the battery to the part, using the clamp to hold the line on.

Make sure the area to be "branded" is clean, bare metal. This means no rust, grease, paint or clear lacquer.

When writing, use a light touch. For the best performance, you'll probably have to make slight bending adjustments to the arm in order to get the right clearance between the bolt and the screw. If the clearance is too large, the electromagnetic force won't pull the screw in. If the clearance is too small, the screw won't have enough room in which to vibrate properly.

Once you're set up, the sparks should fly nicely. The inscription will be rather fine; if you want the letters to be wider, you'll have to do some retracing.

CODED MESSAGES WITH A BUZZER

When Thomas Edison worked as a boy on the Grand Trunk Railroad in Michigan, he became familiar with the telegraph system that linked the eastern part of the United

States with the western part. And, recognizing the vital importance of this cross-country communications system, he later made many improvements to it.

As we said earlier, Edison invented the duplex, the quadruplex and the multiplex telegraph systems, which increased the number of messages that could be sent simultaneously over the line. Before those inventions, a line could handle only one message at a time and in only one direction.

Similar in principle to the initial telegraph system (and, in fact, to the electric pencil) is the buzzer you are about to build. However, instead of producing a single click when the code key is depressed, as did the telegraph, the buzzer gives off a continuous sound.

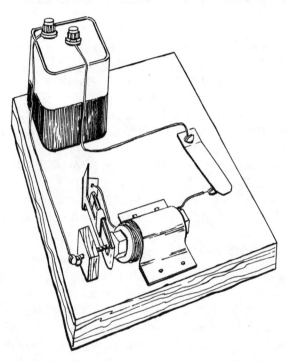

How the buzzer works. For a better understanding of the buzzer's operation, study the circuit diagram. Imagine that you've just pressed the code key. Let's see what happens, starting at the corner terminal of the battery.

Instantly, current shoots down to the brass contactor screw. Since the screw is touching the vibrator arm, the current continues on its way into the coil. Out of the coil, it streaks past the closed code key and back to the battery.

Circuit Diagram

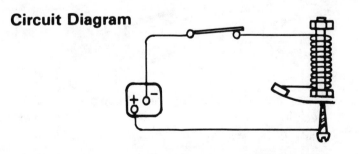

As in the electric pencil, this flow of current creates a magnetic field around the iron bolt. Having become an electromagnet, the bolt attracts the vibrator arm. But as the arm starts to swing toward the bolt, it opens the circuit. Hence the current stops. As a result, the magnetic field collapses, allowing the vibrator arm to swing back against the contactor. With the circuit now restored, current starts flowing again...and the cycle starts anew.

No matter how quickly we press and release the code key, the current will still make hundreds of round trips through the circuit. And because of the resulting rapid motions of the vibrator arm, a buzzing sound is heard.

EXPERIMENT 2: A Code Set

Not only is the code set fun to build, it is even more fun to use, especially with a fellow operator. So that both of you can send, as well as receive messages, you will want to build two identical sets of buzzers and code keys. They're really not hard to make. For each set you will need the following materials.

THINGS NEEDED: ELECTROMAGNET: Bolt and nut $\frac{5}{16}$" by 2"; 2 washers; also the spool of hookup wire used in Experiment No. 1. ELECTROMAGNET COVER: Tin-can strip $1\frac{1}{2}$" by $3\frac{1}{2}$". VIBRATOR ARM: Tin-can strip 1" by 10". SLIDER CLIP FOR VIBRATOR ARM: Tin-can strip $\frac{3}{16}$" by 1". CONTACTOR: Brass screw 1" long. CONTACTOR HOLDER: Wood 1" by 1" by $\frac{3}{8}$". CODE KEY: Tin-can strip $\frac{1}{2}$" by 3". BASE: Wood 7" by 9" by $\frac{3}{4}$". BATTERY: 6 volts (12 volts would be even better). MINOR ITEMS: 6 or 8 small nails, 4 roundhead screws, hammer, file or rasp.

HOW TO BUILD THE PARTS

Winding the electromagnet. Place a washer at each end of the bolt and engage the nut so that it just covers the end of the bolt. Starting at either end of the bolt and leaving about 6" of wire for making connections, begin coiling the wire around the bolt between the washers. Carefully wind one layer along the length of the bolt and another layer back toward the starting point. Keep doing this until several layers of wire have been put on. Plan to finish the winding at the opposite end from which you started. Cut off all but 6" of wire. Now wrap tape around the coil to keep it from unwinding.

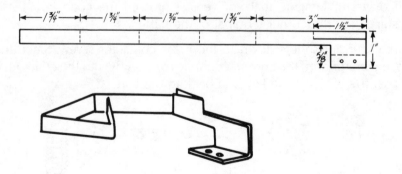

Forming the vibrator arm. Finish cutting and folding the 1" by 10" strip as shown in the drawing. Tap the folded ends with a hammer to keep the layers of metal close together. Punch 2 holes in the base so the arm can be mounted.

Making the contactor. Since you will have to experiment to find the best spacing between the contactor and the bolt head, the contactor should be adjustable. That's why a screw is used. A 1" brass screw with the point filed flat makes a good contactor.

To prepare the holder, lay the 1″ by 1″ by ⅜″ piece of wood flat; and in the exact center, drill a hole slightly smaller than the screw. Drive the screw through the wood.

Cutting out the code key. Once you've snipped a ½″ by 3″ strip from the tin can, the code key is practically made. All that remains to be done is to punch a hole about ½″ from one end so the key can be screwed to the base board. If you want a fancier key, you can attach a small wooden knob to the sending end. With a roundhead screw, fasten the knob to the key by screwing from the bottom up. In that way the screw head can serve as an electrical contact.

MOUNTING THE PARTS ON THE BASE

Using the main drawing as a reference, position the parts accordingly. The electromagnet can be held down in any of a dozen different ways. A simple, yet professional-looking way requires a tin-can strip about 1 ½″ by 3 ½″. Shape the strip around the coil and bend the ends outward so they lie flat on the board. Punch a couple of holes in each end; then nail everything in place.

Next, locate the wooden holder for the contactor about ⅜″ from the bolt end. Secure it in place either by gluing or by nailing from the bottom up.

To install the vibrator arm, proceed as follows: Line up the arm so it is parallel with, and touching, the face of the bolt. Pound a nail partway into the hole that is nearer the bolt. Push the rear of the arm sideways so that the arm swings away from the bolt, causing the thin strip on the other side of the arm to press firmly against the contactor screw. The thin strip and the screw must make full contact with one another. While still pressing against the rear of the arm, put a nail in the remaining hole; then pound both nails all the way in.

Finally, mount the code key, using a screw. Bend the key so that the free end is about ½″ above the board. You will need a contacting terminal on the board directly under the free end of the key. Use a small roundhead screw. If you've put a knob on the key, the base screw should be directly beneath the knob screw. Be sure that the contacting portion under the key is scraped clean to the bare metal.

HOOKING UP THE CIRCUIT

This is the simplest part of the whole project. Just make the connections as indicated on the circuit diagram. In all cases, bare metal should be touching bare metal, and the connections should be tight. It would be best if you could solder the wire leading to the contactor screw. If you can't do this, wrap the wire around the screw and twist it with a pair of pliers. In either case, allow enough wire for the screw to be turned. For the base of the arm, lift one of the nails just enough to tie the wire around it. Then pound it back down.

Both wires to the code key should be looped clockwise around the loosened screws. After tightening the screws and connecting the battery, you are ready to try out the set.

ADJUSTING THE SET

Start with a gap of about ⅛″ between the vibrator arm and the bolt face. Tap the code key a few times. You should get some kind of response. To find the best setting, adjust the contactor screw and the spring force of the vibrator arms against the screw. You can also vary the sound of the buzz with the slider clip. Bend it in half and place it on top of the vibrator arm.

NOTE: Arcing between the contactor screw and the vibrator arm is normal. But it will dirty the contacts, which can in time stop the current. So occasionally clean the screw with a file and scrape the arm with a knife.

CONNECTING TWO SETS

As it now stands, the set can be used by itself. And it will provide hours of fun for anyone interested in learning code. But let's assume that you have a code buddy and that the two of you have made identical sets, each with its own battery. What now?

Interconnection Diagram

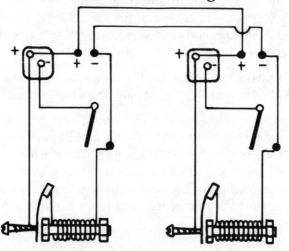

First you will have to add 2 more terminals to each set (see the interconnection diagram). Use screws. Add a wire from the + terminal of the battery to one screw and mark the screw +. Add another wire from the open end of the code key to the other screw and mark that screw −. Upon connecting the 2 sets, run one wire from the + screw and the other wire from − screw to − screw. Now when you tap the code key, your set will work and so will your buddy's. And vice versa. In that way, the sender can hear what he's sending.

USING THE BUZZER

How do we send coded messages with a buzzer? It's fairly easy. The buzzer is a form of doorbell, the kind that rings steadily as long as the button is held in. By depressing the buzzer's code key for a split second, we produce a short sound. And by holding the key down a trifle longer, we produce a longer sound.

These short and long sounds can be combined in different ways to represent letters of the alphabet. The Morse code, sound language used by radio operators all over the world, tells us what combination of short and long sounds (dots and dashes) stands for each letter.

Actually, you can "talk" Morse code as well as send it by key. For example, in pronouncing the letter a (.-), an experienced radio operator would say "di dah." For the

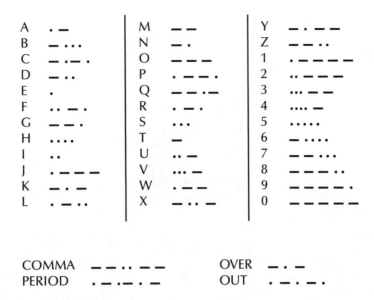

A	. —	M	— —	Y	— . — —
B	— . . .	N	— .	Z	— — . .
C	— . — .	O	— — —	1	. — — — —
D	— . .	P	. — — .	2	. . — — —
E	.	Q	— — . —	3	. . . — —
F	. . — .	R	. — .	4	. . . . —
G	— — .	S	. . .	5	. . . . .
H	. . . .	T	—	6	— . . . .
I	. .	U	. . —	7	— — . . .
J	. — — —	V	. . . —	8	— — — . .
K	— . —	W	. — —	9	— — — — .
L	. — . .	X	— . . —	0	— — — — —

COMMA	— — . . — —	OVER	— . —
PERIOD	. — . — . —	OUT	. — . — .

letter b(—...), he would say "dah di di dit." Knowing the sounds of each letter in this manner helps in learning the code.

When sending a message, allow a little time between letters and more time between words. The message "good night" should go like this:

G O O D
—. ——— ——— —..

N I G H T
—. .. ——. .... —

If the sender of this message wanted a reply, he would wait a moment and tap the letter k(—.—). This is the code talk for "over." In effect, it means, "It's your turn to send." If the sender expected no reply, he would tap out the letters ar (.—.—), without a pause between the letters. This is also code for "out." And this, in effect, means, "I won't be sending or receiving anymore, for now."

LIGHT FROM A "FRANKENSTEIN" BATTERY

Although his greatest achievements marked him as an inventor of electrical and mechanical things, Thomas Edison always thought of himself as a chemist. "Grand science, chemistry," he said. "I like it best of all the sciences."

Without question, Edison's most outstanding invention in the field of chemistry was the nickel-iron-alkaline battery. He perfected this revolutionary battery in 1908, after ten years of research, 50,000 experiments and a million dollars of his own money. His relentless hunt for a good storage battery has become a famous chapter in the history of applied science. Equally remarkable, his nickel-iron-alkaline battery, although created half a century ago, is still being made and used today—in this era of space probes, lasers and nuclear power plants.

We couldn't hope to build even a crude version of Edison's battery. For one thing, it requires special materials. But we can make a much simpler type of battery. By doing so, we will get a clearer picture of what a battery is and why it delivers current.

Actually, we're going to make a battery out of the basic materials used in most flashlight batteries. In fact, we're going to start with the insides of a dead battery and sort of bring it back to life. This "Frankenstein" battery will even be able to do some work for us.

HOW OUR LAMP WORKS

Before going any farther, we should point out a slight but very common mistake that most of us make in using the word "battery." The so-called flashlight battery, for example, is really not a battery. It is a cell, commonly known as a dry cell (although it does contain a moist mixture). On the other hand, the automobile storage battery is a battery. That's because it houses several cells (wet cells, in this case) that are connected internally. A battery, then, is a group of two or more cells, wet or dry, connected as one unit. But habits are hard to break, and we still find ourselves calling the cell a battery.

All electrical cells consist of an electrolyte (a conducting solution or paste) and two electrodes. The wet-cell lamp you're going to make uses ammonium chloride, also

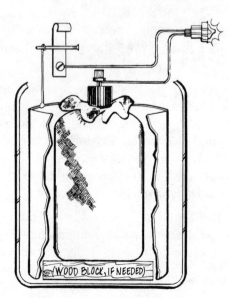

called sal ammoniac, as the electrolyte. Zinc is one electrode and carbon (including the black mixture around the carbon) is the other electrode.

Look at the circuit diagram. When the lamp is operating, the zinc slowly dissolves in the electrolyte. As the dissolving goes on, an excess of electrons builds up in the zinc. Since all electrons have the same electrical charge, they repel one another along the circuit. Thus, they force their way into the bulb filament, pass through the carbon rod and enter the black mixture, where other reactions occur. It is this continuing movement of electrons along the circuit that we call current and that causes the high-resistance filament to become white-hot.

EXPERIMENT 3: A Simple Battery

THINGS NEEDED: CARBON ELECTRODE: A worn-out #6 dry cell (if you plan to build the cigar-box microphone described later, try to get 2 dry cells while you're at it). ZINC ELECTRODE: Zinc sheet about 5" by 10" (use whatever zinc you can salvage from the dry cell; or try galvanized sheet metal, which is zinc-coated). ELECTROLYTE: Ammonium chloride (get a small block of sal ammoniac from the hardware store; it's used for cleaning soldering irons and is quite inexpensive). CONTAINER: A wide-mouthed quart jar, including lid and waxed cardboard liner. LAMP BASE: Wood disk about 4" in diameter. LIGHT SOURCE: A 1-cell penlight bulb with socket. SOCKET SUPPORT: Tin-can strip 1" by 2". SWITCH: Tin-can strip ¼" by 2". MINOR ITEMS: Hookup wire, aluminum foil, napkin, old nylon stocking, small nails, small screw, wax (optional), pencil, ruler, string, metal saw, drill.

HOW TO PREPARE THE PARTS

The carbon electrode. Let's begin by opening up the old dry cell. We should try to do it in a way that won't disturb the caked mixture around the carbon rod. Measure upward ½" from the bottom of the cell and draw a line around the outside. With a metal saw, make a series of cuts through the zinc casing, turning the cell slightly after each cut. When you have finished, you will be able to pull the bottom off. Then do the same thing at the top of the cell, but this time measure downward about 1¼". After you have cut through the casing at the top, run the saw into the mixture until you reach the carbon rod. A little twisting now will enable you to remove the top.

At this point, start sawing the casing lengthwise. When through the casing, spread it open...and there's our electrode. The black mixture should still be in one piece around the rod, and about 1" of rod should be exposed at the top. Carefully wrap a napkin or paper towel around the electrode so that it covers all the mixture. Then slip the wrapped electrode into a nylon stocking. Tie some string around the stocking above the mixture and cut off the rest of the stocking. That takes care of the carbon electrode. (The napkin and stocking simply keep the mixture together.)

The zinc electrode. If you've managed to get a piece of zinc about 5" by 10", form it into a cylinder that just fits the container you plan to use. The top of the zinc should be about 2" below the top of the container. Also, the zinc should have about a 6" length of copper wire soldered to it. If you don't have soldering equipment, pierce a hole in the zinc, thread the copper wire through it and loop the wire around itself tightly.

The electrolyte. You're probably wondering how that rock-hard sal-ammoniac block is ever going to dissolve so we can get our electrolyte. The best way to handle this situation is to wrap the block in an old but sturdy rag. Then lay the block on cement and break it up with a hammer. The smaller the pieces, the quicker the dissolving. Then unwrap the rag and empty the contents into a container other than the one intended for the battery. Add a cup of warm water, stir with a stick and set the container aside for the time being.

The lid and lamp base. What we have to do here is to form the jar lid and wood disk into one piece. So first carefully remove the waxed liner from the lid. Also remove from the center of the lid a circle of metal about 1 ½" in diameter. This doesn't have to be done accurately, or even neatly. Any curled edges should be flattened with a hammer.

Next, make a 1"-diameter hole in the center of the wood disk. The hole should be big enough to allow the carbon rod to fit easily into it. Now, with the wood disk placed on something you can hammer on, center the lid on the disk and fasten it with a few small nails. Pound the nails right through the lid into the disk. The nails may go through the disk, so make sure you are not working on a good surface.

Either file the protruding nail points flat or simply bend them over. Replace the waxed liner and cut out the 1" hole with a knife (careful now). Finally, about halfway between the inner and outer edges of the disk, drill a ⅛" hole so that the wire from the zinc electrode can be brought to the top.

You can skip the next step if you want to. But ammonium chloride, as it crystallizes, may creep up the inside wall of the container or up the carbon rod and eventually work its way to the disk. To control this creeping, let the wax from a burning candle drip around the inside lip of the container and around the carbon rod (except for the part that fits into the hole in the disk).

Light source and switch. For the socket support, make a 90-degree bend at one end of the 1" by 2" tin-can strip; at the other end, form a hole big enough to hold the socket snugly. Nail this support to the wood as shown.

Before inserting the socket, take a sheet of aluminum foil and fold it in half about 3 times to make it stiff. Form it into a miniature reflector, poke a hole in the center and slip it into the socket. Then insert the socket into the support.

The switch (the ¼" by 2" tin-can strip) makes contact with a bent nail. Screw it to the disk wherever convenient.

ASSEMBLING THE CELL-LAMP

Place the carbon electrode in the container temporarily to see if the top of the carbon will be high enough. If not, you'll have to raise it by putting a block of wood on the bottom of the container. Now fill the container about half full of the electrolyte. Slip the zinc cylinder over the carbon electrode and lower the two into the container. The electrolyte should cover the zinc.

Through the small hole, feed the wire from the zinc electrode. Then screw the lid on the jar firmly. Tap a small nail into the disk next to the switch. After hooking the zinc electrode wire around the base of the nail, bend the nail over so that the switch slides under it. Connect one of the socket wires to the screw on the switch; run the other socket wire to the carbon electrode terminal.

Now we're in business. Flip the switch and our Frankenstein cell-lamp springs to life instantly. It should last for hours. If you have any trouble try a new bulb, or check the zinc electrode wire; it may have come off. You may also try to agitate the jar a few times or jiggle the carbon rod a little.

A RADIO THAT PLAYS FOR FREE

If someone asked you to name the man who invented radio, you wouldn't be able to answer very easily. That's because a great many men did something important toward making the radio possible. And even though no single person deserves the most credit, certainly among the major contributors was Thomas Alva Edison.

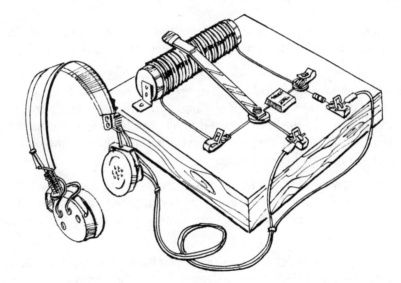

Edison, in fact, made the world's first electronic tube, in 1880. Scientists in those days named it the Edison effect lamp. It consisted of an incandescent lamp into which Edison sealed a small metal plate opposite the filament. The only conducting material the plate touched was a wire leading outside the bulb to a current indicator.

Edison built this device because he was trying to learn why the insides of incandescent lamps developed a dark coating after a while. He thought that maybe current existed in the space within the horseshoe-shaped carbon filament and might be carrying carbon particles to the glass. When he started experimenting, he found that every time the lamp was turned on, the indicator registered a reading. The brighter the light, the higher the reading.

This meant that current was moving from the filament to the plate by traveling through space...in other words, without wires (which is what the word electronic implies). Although he didn't realize it, Edison had built the basic radio tube. His patent (No. 307,031) on the Edison effect lamp eventually became the cornerstone of the electronics industry.

ETHERIC FORCE: RADIO WAVES

But the tube wasn't Edison's only contribution to radio. He made another important discovery. This one concerned electrical energy radiation. In working with an electromagnetic vibrator (similar in principle to the common doorbell), he found that he could draw sparks from the vibrating arm by touching it with a wire. The sparks didn't behave like ordinary electrical sparks though. For example, they wouldn't charge an electroscope. After much study and experimentation, Edison concluded that the sparks represented a "true unknown force." He referred to it as "etheric force." In reality, Edison had been experimenting with what are presently known as electromagnetic, or radio, waves. And his etheroscope (for detecting etheric force) thus became the first detector of such waves.

Now that we know a little about Edison's contributions to radio, let's turn our attention to the radio itself. One of the questions that may be popping into your mind at this point is...

HOW DOES A RADIO WORK?

Radio principles are not the easiest things in the world to understand. So let's look at a general explanation of what is happening as you listen to your crystal radio.

Suppose the weatherman is talking. The vocal sounds he makes into the microphone at the broadcasting studio are converted into electrical signals. After going through various stages of electronic hocus-pocus, the treated signals are fed into the transmitting antenna. There they are radiated in all directions as waves of a frequency belonging to that station (whose frequency is different from that of any other station in your area; otherwise, you'd hear all stations at once).

As the incoming waves cut across the antenna of your crystal receiver, they induce signals of that station's frequency in the antenna. The induced signals enter the receiver, which converts them back to sounds that are almost identical to those made by the weatherman.

When we change the slider-arm position, we change the receiver's sensitivity to the frequency of the station we were listening to. At the same time, however, the moving slider arm makes the receiver sensitive to other frequencies. And we can now pick up stations broadcasting on those frequencies...provided, of course, that these stations are in the area and transmit a fairly strong signal.

EXPERIMENT 4: A Basic Radio

THINGS NEEDED: TUNING COIL: Spool of #16 magnet wire. CORE: Wood dowel 1" diameter by 5". SLIDER ARM: Stiff piece of metal about 5" by ⅜". BASE: Wood 8" by 8" by ¾". CAPACITOR: Mica capacitor 0.002 mfd. CRYSTAL DETECTOR: Germanium diode, IN34A. ANTENNA: Wire 50' to 100' long. (Use whatever kind you have on hand, bare or insulated. If bare, you'll need a glass insulator at each end of the antenna.) HEADPHONES: A high-impedance pair of 2000 ohms (the common transistor-radio earphone may not work too well, if at all). MINOR ITEMS: 4 Fahnestock spring clips, 5 screws, 2 washers, 6 small nails, 2 tin-can strips 1" by 1 ½", hookup wire, drill, soldering iron or gun, saw, sandpaper.

HOW TO BUILD IT

Tuning coil and core. If you can't find any 1"-wood doweling for the core, there are other things that will work as well...maybe even better. A piece of 1"-outer-diameter rubber hose would be excellent, so would a stiff plastic tube or rod. Or take a piece of "one by two" wood, which actually measures ¾" by 1½", and saw it lengthwise. This will give you a piece ¾" by ¾".

After selecting your core and cutting it to the 5" length, measure off ½" from each end. Then drill (or pierce) a small hole through the core at both marks. These holes will keep the magnet wire in place. Thread about 4" of wire into one of the holes and begin winding the coil. Keep the turns close to one another. It will take about 70 turns to reach the other hole. When you reach it, thread the wire through and cut off all but 4". Assuming you've used wood as the core, make supports for it with the 1" by 1½" tin-can strips folded lengthwise for stiffness. Mount the tuning coil and core as shown in the main drawing. If you've used tubing as the core, simply lay it flat and nail it in place.

The slider arm. Bend the 5" by ⅜" metal strip as shown in the sketch. A thick piece of bare copper wire will have to be soldered to the underside of the front. This wire allows the arm to make contact with no more than 1 or 2 turns of wire on the coil. That's important for good tuning. You can use a piece of magnet wire for this purpose if you scrape the clear insulating coating from it. Finish the slider by wrapping tape around the front edge. The tape prevents your touching the bare metal, which could weaken the signals.

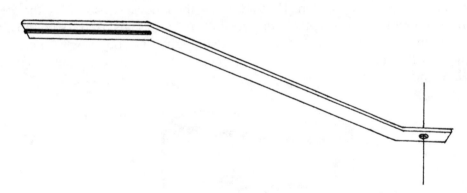

ASSEMBLING THE SET

Refer once again to the main drawing, or to the circuit diagram if you prefer, and start putting the parts together. It doesn't make any difference which way you insert the capacitor. The same goes for the crystal diode. If possible, make as many connections as you can by soldering. Weak signals can be lost through poor connections.

When installing the slider, be certain that the copper wire underneath touches the coil throughout the full swing of the arm. Also, take a piece of sandpaper and remove all the insulation from the top of the coil...right down to the bare metal. The slider must be able to make contact with each turn of wire on the coil.

To get the most out of your crystal set, you must have a good ground connection and a good antenna. Cold-water pipes make excellent conductors to ground. Make sure that the pipe has been sanded or scraped clean at the point you plan to make the connection.

As a starter, string up a temporary antenna to see what your crystal set can do. Use about 100' of wire if you can, and locate it as high as is practical. But don't run the wire

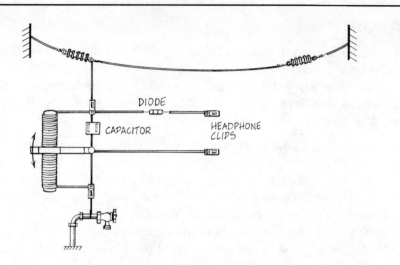

under or near power lines or leave it up when you're not using the set (and never use the set during a storm). Should you decide to erect a permanent antenna, you'd better use a lightning arrester and get an expert's advice on installing the antenna properly.

OPERATING THE SET

Now, with everything assembled tightly, the ground and antenna wires connected, and headphones clipped on, we're ready for the big moment. Move the slider slowly until you pick up a station. Then adjust for the loudest sound. Since the set we have built is a rather simple one, it won't receive many stations. And it's possible that more than one station will come in at the same time. But the sounds will be clear and thrilling, and the set won't cost a penny to operate. So, happy experimenting...and good listening.

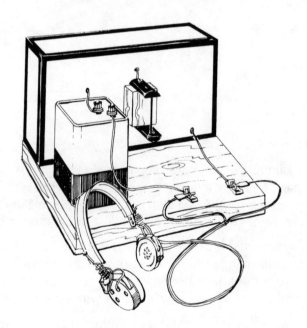

A SUPERSENSITIVE CIGAR-BOX MICROPHONE

Although Alexander Graham Bell is credited with the invention of the telephone, it was Thomas Edison who devised the first telephone transmitter that could be used over long distances. Unlike Bell's limited-range instrument, Edison's transmitter took advantage of a wonderful property of carbon: If a loose pack of carbon particles is squeezed, the electrical resistance of the pack decreases. In other words, when current is passing through the pack, additional current will flow when pressure is applied.

Edison had the idea that voice waves could apply that pressure. He was right. In the carbon transmitter he perfected, loud voice sounds (upon striking the carbon particles and compressing them) caused larger currents than did quieter sounds. These variations in current traveling down a transmission line regulated a receiver at the other end of the line, and that receiver reproduced the sounds of the speaker's voice. This use of carbon in a telephone is still practiced today. Edison's carbon-particle device, then, was the forerunner of the modern telephone transmitter and the microphone used in radio broadcasting.

The cigar-box microphone, shown here, is similar to Edison's in at least one respect: It is a closed-circuit system, which means that current is constantly flowing. Edison's first "speaking telegraph transmitter" (Patent No. 474,320) included this important concept. Bell's instrument did not, which is one of the reasons its range was limited to only a few miles. However, the cigar-box "mike" is not a carbon-particle transmitter, even though it uses carbon. It is a loose-contact mike. It won't give anywhere near the sound quality that Edison's did. Nevertheless, it is an extremely sensitive detector of sound and one that can be fun to make and to experiment with.

EXPERIMENT 5: A Cigar-box Microphone

Being a loose-contact detector, the cigar-box microphone has the same high sensitivity to vibrations as do insecure electrical connections. You've no doubt noticed how easily a loose light bulb flickers when someone passes by. So it is with our mike (see the circuit diagram). The carbon electrodes loosely support the pencil-lead rod. The slightest vibration, like from a sound, will disturb the rod. When the circuit is closed and current is flowing through the headphones, this disturbance changes the current flow. The headphones respond to these changes and, hence, tend to imitate the sound.

THINGS NEEDED: CARBON ELECTRODES: The carbon electrode from an old #6 dry cell. CARBON ROD: 2" length of lead from a wooden pencil. ELECTRODE BASE: Wood 2" long by ¾" thick (the width doesn't matter). SOUNDING BOARD: Cigar box (the cover isn't needed). MIKE BASE: Wood 7" by 9" by ¾". BATTERY: 6 volts.

HEADPHONES: The same ones used for the crystal radio experiment. MINOR ITEMS: 6 screws, 4 nails, 2 Fahnestock spring clips, hookup wire, drill, file or fine sandpaper.

HOW TO PREPARE THE PARTS

Forming the electrodes. Start by removing the 1"-diameter carbon electrode from the dry cell. Break it open with a hammer and chisel. Clean the bottom of the electrode and saw off a 1½" segment. Then saw the segment in half lengthwise, giving us 2 half-cylinder electrodes. You might want to file or sand the rough surfaces a bit.

To screw the electrodes to the wood base, which is ¾" thick, we'll have to drill a hole in each electrode about ⅜" from one of the ends. Next we'll have to make two small depressions for the pencil lead to rest in. Do this by drilling a dimple in each electrode about ¼" from the end opposite the screw hole.

The loose-contact assembly. Lay the wood base on a flat surface and stand the electrodes at the ends, directly facing one another. Then secure the electrodes to the base with screws. These screws will also serve as wire contacts.

Now here's the tricky part. The pencil lead must fit between the tightened electrodes so that it is free to move slightly. If it is held firm, our loose-contact microphone won't have a loose contact...consequently, it won't work. At the same time, the fit shouldn't be sloppy. So do the best you can. Before you install the pencil lead, be sure to sharpen both ends on a file or fine sandpaper.

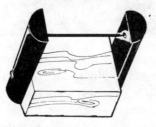

PUTTING THE MICROPHONE TOGETHER

Position the loose-contact assembly on the cigar box as shown in the main drawing. Fasten the assembly to the box with 2 screws coming from inside the box. Also, fasten the cigar box to the base as shown, using the 4 nails. The rear of the box should extend slightly beyond the edge of the base.

The Fahnestock clips, which accommodate headphones with separated terminals, may be installed wherever you like. Possibly your headphone set (or single headphone) has a phone-plug connector. In that event, you'll have to obtain a suitable receptacle.

With a pointed tool, poke 4 holes in the cigar box, as illustrated, to pass the wires through. This operation isn't necessary, as you can see, since the microphone can be completely wired at the front of the box. However, it does result in a neater-looking job. Assuming that you've made the holes, run a wire from the upper electrode screw into the box and then back out of the box to the center terminal of the battery. Do the same at the lower electrode, except run the wire to the screw on the outer Fahnestock clip. Conclude the hookup by linking the inner clip to the corner battery terminal. Now let's put our workmanship to the test and see how well the mike operates.

USING THE MIKE

Connecting your headphones to the microphone completes the circuit and turns the mike on. You'll be amazed at the new world of sound opened up to you. You should be able to hear your own breath blown against the pencil lead. Grains of salt dropped on the electrodes should sound like rocks; tapping the box with your fingernail might pass for a mild explosion. Try placing a spring-wound alarm clock or a wristwatch on the box. Also lean a transistor radio against the box; even with the volume set at low, you'll be able to hear the program through the headphones. The radio experiment will be even more impressive if you can put the mike in one room and the headphones in another.

What else can we do with the mike? Well, how about using a sewing needle instead of the pencil lead? Try it with the point down, then with the point up; turn it to find spots of higher sensitivity. Lay the box flat. Hold it against a wall and have someone talk on the other side of the wall. Also, see what happens when you adjust the electrodes, change pencil-lead hardness, or substitute a flashlight battery for the 6-volt battery.

LOOKING AHEAD INTO THE FUTURE

A little over a hundred years ago, with Edison's invention of electric light spreading rapidly across the nation and from urban to rural communities, the thought of energy conservation was of concern to no one. Growth and expansion were the buzz words of the day.

Times, however, have changed. Today we realize that our present sources of energy are indeed limited. Each source has problems associated with its usage.

Were Thomas Edison alive during the oil embargo and witness to the energy crisis, all of his efforts would have gone into conservation research and a search for new and better sources of energy. We need alternatives and more efficient utilization of Planet Earth's raw materials in the form of coal and other fossil fuels. We need a modern-day Thomas Edison.

More likely, if we are to solve the energy problem, we need many Edisons. In the words of the late Charles Franklin "Boss" Kettering, "One Thomas Edison in a generation is no longer enough. Today, if we are to sustain the present rate of progress, we need countless thousands of men with the wisdom, imagination and perseverance of Thomas Edison."

The problems that confront mankind today and those we will face in the future are far more complex than those faced by Edison. Maybe it was his realization of this that led him, when asked what he considered his greatest invention, to answer without hesitation: "The commercial laboratory."

Edison realized that with the increasing complexities of science, the day of the lone inventor was waning. He was the first to systematically organize experimentation, the backbone behind all of his inventions.

Today most of the world's inventions come from commercial laboratories, where, thanks to modern technology, experimentation is reaching new levels of organization and speed. Much of the emphasis in modern-day "Edison laboratories" is focused on new sources of generating the energy that Edison's inventions made so popular.

The projects in the two following sections, "Energy for the Future" and "Alternative Energy Sources," will show you how you can apply the same basic method used in the world's greatest laboratories, the method Edison relied upon so heavily—that is, the experiment—to learn more about how to conserve energy and what new sources may someday be utilized to generate it.

Edison at his ore-processing plant in Ogdensburg, New Jersey, in the mid-1890s. The plant was built to implement his invented process for separating low-grade iron ore from the worthless material found with it.

No small operation, the plant had miles of conveyors, giant rollers that could crush piano-sized ore-bearing rocks, and the largest steam shovel in America.

PART FIVE

ENERGY FOR THE FUTURE

BEFORE YOU BEGIN

Before you begin the experiments in this section, take a minute to think about energy. What is it? Where does it come from? What is it used for? And why are people all over the world so concerned about the "energy crisis?"

In simple terms, energy is the ability to do work. There are many forms of energy: radiant (which includes light), thermal, chemical, mechanical, electrical, nuclear and gravitational. Energy itself is never consumed (that's the "First Law of Thermodynamics"); it's only changed from one of these forms to another.

The problem is, whenever energy changes form, some of it is lost. It changes into unusable heat, which dissipates and gradually warms up the earth's atmosphere (and that's the "Second Law of Thermodynamics").

As an example, look at all the changes energy could undergo in order to light your living room tonight (see picture).

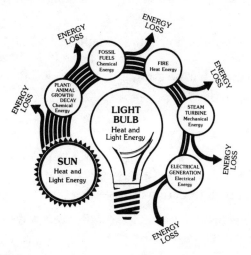

CONVERSIONS MEAN LOST ENERGY

The more conversions there are between the primary source of energy and its final use, the greater the waste. In addition, some machines simply use more energy than other machines performing the same task. In 1979, the United States wasted more energy than it had both used and wasted in 1960. This is one reason behind the energy crisis.

Another reason is our dependence on fossil fuels, especially oil and natural gas, as our primary supply of energy. These resources were created millions of years ago. Unlike sunlight, wind, water or living plants, they cannot be replaced. You can grow another tree, but once a barrel of oil is gone, it's gone forever.

So now the race is on to find substitutes for oil and natural gas before these sources dry up or become so scarce that few people can afford them. It's a race against time. Some experts predict, for example, that the United States oil reserves will be used up by about the year 2035. They're the optimists. The pessimists are predicting 1998!

This brings us to energy conservation today. Every bit of oil or gas energy that we save now will "buy" us that much more time in which to come up with other sources to generate electricity, power our transportation systems and produce the many products we have come to take for granted.

WHAT CAN WE DO TO HELP?

These pages will give you a better understanding of how energy is lost. You will take a look at some of the factors affecting your own family's energy consumption; you will investigate some of the ways to reduce that energy consumption. And, finally, we will talk about solar energy—one of the most exciting and versatile alternatives to oil and natural gas.

We hope you will have fun doing the experiments...and that you will learn something too, of course. But more than that, we hope these exercises will raise your "energy awareness" so that you will be constantly on the alert for ways to save, or avoid wasting, energy in your daily life.

GETTING THE FACTS

The first step in deciding how to conserve energy is to identify some of the major factors that cause you to lose—or help you to save—energy in your home. Only then can you develop a logical plan to cut your losses.

Here are two experiments that will help you discover ways in which you may be losing energy right now in your own house.

EXPERIMENT 1: Your Home and the Forces of Nature

THINGS NEEDED: Compass and pencil.

There is not much you can do about your home's location, especially since moving houses is an expensive and tricky proposition. But knowing how that location affects energy usage can be valuable when you're beginning your energy audit in the next experiment.

First determine the orientation of your house and how the weather forces act upon it, as shown on page 74. Using the compass, locate the points at which the sun rises and sets. You can also obtain information about the directions of the winds and storms

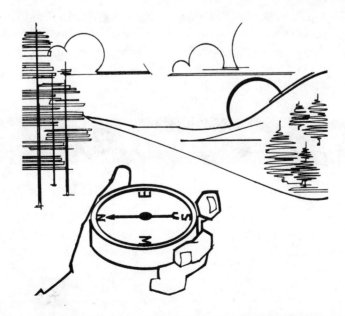

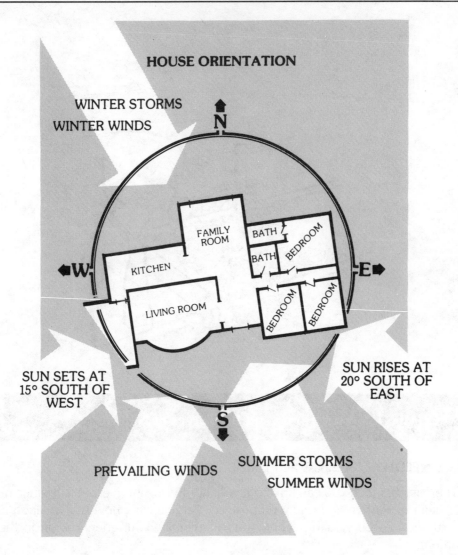

HOUSE ORIENTATION

WINTER STORMS
WINTER WINDS

N

W

E

FAMILY
ROOM

BATH

BATH

BEDROOM

KITCHEN

LIVING ROOM

BEDROOM

BEDROOM

SUN SETS AT
15° SOUTH OF
WEST

SUN RISES AT
20° SOUTH OF
EAST

S

PREVAILING WINDS

SUMMER STORMS

SUMMER WINDS

that affect your area by calling your local weather bureau, the state agricultural extension service or news agencies. Then sketch in the areas of your house that are used most often—the kitchen, dining room and family room, for example.

Now take a few minutes to think about your house in relation to the forces of nature. If your house is in a northern area, for instance, shrubbery along the northern-most wall might give it added protection from winter storms. Are there drapes or curtains in front of windows exposed to the winter winds? Could anything be done to your home to give it more protection?

EXPERIMENT 2: A Basic Home-energy Audit

THINGS NEEDED: Pad, pencil, flashlight, gloves, ruler, draft detector (see picture on Page 76), 2 thermometers and tape.

First find your geographical location on the map below to determine the approximate inches of ceiling insulation recommended for your temperature zone.

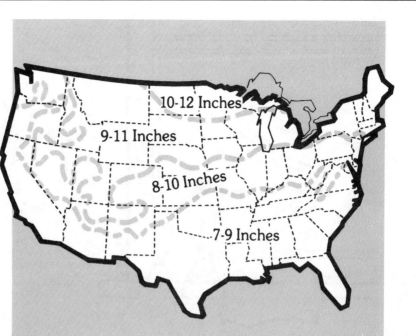

In The Attic

• Measure the depth of insulation between ceiling joists. (Remember to wear gloves when you handle insulation.)

• If you already have the recommended thickness of insulation or more, score 30 points. If you have 2″ less, score 25; 4″ less, score 15, 6″ less, score 5. If you have less than 2″ of insulation, score 0.

In the Living Areas

To investigate these areas, you will need a draft detector. So...

QUICKIE EXPERIMENT "A"

Making a Draft Detector

Cut out a piece of clear plastic food wrap about 5″ × 10″. Tape the 5″ end to a short stick so the long end hangs freely. For the stick, you could use a pencil or ruler (but a baseball bat probably wouldn't be a good idea.)

That's your draft detector. Notice that even when you blow ever so lightly, the plastic detects the movement of air.

• Check for drafts by holding the draft detector about 1" from the place where windows and doors meet their frames.

    If there is no draft around your windows, score 10 points.

    If there is a draft, score 0.   _____

    If there is no draft around your doors, score 5 points. If there is a draft, score 0.

                                                                                               _____

• If you live in an area where the temperature often falls below 30° F and you have storm windows, score 20 points. If you do not have storm windows, score 0. _____

• Test your wall insulation by taping one thermometer to an outside wall and taping another in the center of the same room (say, to a chair back). Make sure that the thermometers are at the same height from the floor. Read the thermometers after 4 hours.

    If there's less than 5 degrees difference, score 10 points. If the outside-wall reading is more than 5 degrees below the inside reading, score 0. _____

• Do you have a fireplace? If not, add 4 points. If you do have a fireplace and always keep the damper closed when it is not being used, add 4; if you leave the damper open when the fireplace is not in use, score 0. _____

    Now, time for a little test.

QUICKIE EXPERIMENT "B"
That Hole in the Roof
For a dramatic demonstration of how much heat is lost up the chimney in the winter, check the fireplace with your draft detector. Start with the damper open, then with it closed. You'll be amazed. The escaping heat is like money going up in smoke.

In the Basement

• If you have a heated basement or if there is no space under the house, score 10 points. If you have an unheated space under the house and there is insulation under your floor, score 10; if there is no insulation, score 0. _____

YOUR ENERGY CONSERVATION HABITS

• In winter, if your thermostat is set at 68° F or less during the day, score 6 points. Deduct one point for every degree over 68. If your thermostat is set over 70°, score 0. _____

• If you set your thermostat back to 60° F or less at night during the winter, score 10 points. Deduct one point for each degree over 60. If your thermostat is set at 66° F or above, score 0. _____

• If you do not have air conditioning, score 7 points. If you have air conditioning and you keep the temperature setting at 78° F or above in summer, score 5. Deduct one point for each degree below 78. If your air conditioning is set below 76° F score 0. _____

• IF your water heater is set below 120° F, score 10 points. If it is set between 120° F and 140° F, score 5. And if it is set above 140° F, score 0. _____

TOTAL _____

NOTE: IF YOUR SCORE ON THIS HOME-ENERGY AUDIT IS LESS THAN 100, THE MEMBERS OF YOUR HOUSEHOLD PROBABLY COULD TAKE SOME STEPS TO MAKE THE HOUSE MORE ENERGY-EFFICIENT—AND SAVE SOME MONEY IN THE PROCESS. WHY NOT HAVE EVERYONE MAKE SUGGESTIONS AND THEN LET THE GROUP VOTE ON THEM?

CONSERVING ENERGY

Now that you know how you are using energy in your home, you can begin looking for ways to cut down on the amount you use.

Try this for a start: Simply use more wisely some of the things in your home that are run by electricity. Or, when considering a purchase of an electrical product, compare the energy-saving features of various manufacturers' makes and models.

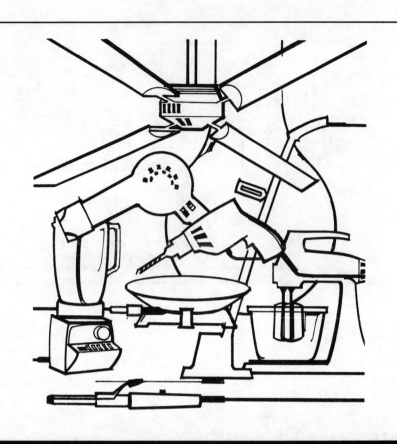

EXPERIMENT 3: Can You Use Electricity More Wisely?

THINGS NEEDED: Pad and pencil, and cooperation from everyone in the house.

Get together with the rest of the family and go over the list of common electrical appliances listed below (of course, just the ones you have in your house); add anything pertinent you may have that is not on the list.

Which appliance could you use less often and still be comfortable? Put a check in the column on the chart that best describes your family's attitude about each appliance. A simple majority could determine the family vote.

Electrical Appliances We Can Use More Wisely (cont'd.)

	Very Easily	Easily	With Some Difficulty	With Great Difficulty	Impossible
Air conditioner					
Blanket(s)					
Blender					
Calculator					
Can opener					
Carving knife					
Clock(s)					

Electrical Appliances We Can Use More Wisely

	Very Easily	Easily	With Some Difficulty	With Great Difficulty	Impossible
Clothes dryer					
Clothes washer					
Curling iron					
Dishwasher					
Fan(s)					
Freezer					
Frying pan(s)					
Game(s)					
Garbage disposal					
Hair dryer(s)					
Hot water heater					
Ice crusher					
Ice cream maker					
Iron					
Lights					
Microwave oven					
Mixer					
Popcorn machine					
Power tool(s)					
Radio(s)					
Range					
Refrigerator					
Space heater(s)					
Stereo/Record player(s)					
Television(s)					
Toaster					
Toothbrush(s)					
Trash compactor					
Typewriter					
Vacuum cleaner					

Agreeing on the list was the hard part; the rest of the experiment is less difficult.

First you should know how to read an electric meter. An electric meter measures the number of kilowatt-hours (KWH) consumed. Most electric meters have four or five dials, each numbered from 0 to 9. To read the dials, write down the number the pointer has just passed. Be careful to note whether the dial reads clockwise or counter-clockwise.

Look at the examples shown below and make sure you understand how the readings were made. (Four-dial meters do not measure individual KWH units, so you will have to add a 0 as the last digit in your reading).

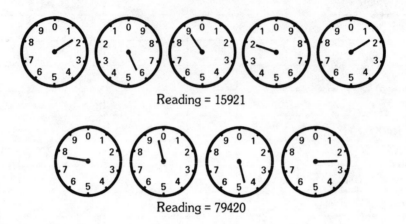

Reading = 15921

Reading = 79420

Keep a chart like the one below to measure your electricity usage. To determine how much electricity your family uses during a normal week, subtract the reading you take at the beginning of the week from the reading you get at the end.

	Reading (KWH)	Use (KWH)
Reading on *October 31*	*37,271*	*—*
November 7	*37,414*	*143*

Now for the energy-saving experiment. Begin by taking a meter reading. Then have the family avoid using all the appliances in the "Very Easily" column. After a week, read the meter again and enter the reading on the chart. Do you see a big drop over a normal week's usage? Probably not. The appliances in this column are likely to be the ones that don't get used much anyway.

So for the next week, ask the family to avoid using appliances in both the "Very Easily" and the "Easily" columns. Take your meter reading at the end of the week, log it, and see if there's a drop. And for the third week, ask the family not to use the items from the first three columns of the list, and so on.

By the end of the experiment, you should see a slight reduction in the amount of energy used per week. This assumes that the weather stays about the same and that the energy used by the appliances still plugged in did not change drastically during the experiment.

While running this experiment, remember to make allowances for other factors. During holidays, for example, major appliances tend to be used a lot more than usual—or a lot less. And a sudden cold snap or a hot spell could mean that any savings were hidden by the increased use of your central heating system or air-conditioning unit. Even normal variations in usage may hide real reductions. So don't be discouraged if your tests show smaller savings than you expected. Think of how much higher your readings would have been if all the appliances were being used.

After this experiment, your family may decide to test other energy-conservation measures (like turning down the room thermostat or the setting on the hot-water tank) to see how they affect energy consumption.

Before we leave the subject of using less energy in the home, here's a simple demonstration of how to conserve heat when cooking:

QUICKIE EXPERIMENT "C"
A Cover Is Like a Blanket

Pour two cups of water in a pan and set the pan on a hot range burner. Note how long it takes the water to boil vigorously.

Empty the pan and let it cool. Then repeat the steps above. Only this time, put a cover on the pan. You'll know the water is bubbling away when the cover starts rattling.

Did the covered water come to a boil faster? Is it worth using a cover when boiling? Now you know.

EXPERIMENT 4: Building and Weatherizing a Model House

THINGS NEEDED: One 4' by 8' sheet of ¼" plywood, finished on one side; about 100' of 1" by 2" furring strip; one pound of 1½" nails; about 9 square feet of foam padding ¾" thick for insulation; a small bag of loose insulation; a 9' by 12' heavy-duty, clear-plastic drop cloth; thumbtacks; glue; tape; 8' of adhesive-backed weatherstripping ¾" wide; saw, hammer and pair of pliers.

Your home loses heat in winter (or gains it in summer) in two main ways. Conduction occurs when heat moves directly through the ceilings, walls and windows from the warmer to the cooler side.

Insulation makes a big difference in fighting conduction. Infiltration occurs when cold air leaks in while hot air leaks out (or vice versa) through cracks around doors and windows, or in the foundation or siding. Weatherstripping and caulking, especially around doors and windows, stop excessive infiltration.

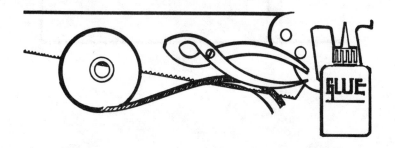

CUTOUT PATTERN FOR 4' x 8' PLYWOOD SHEET
(Cut Each Piece as Needed)

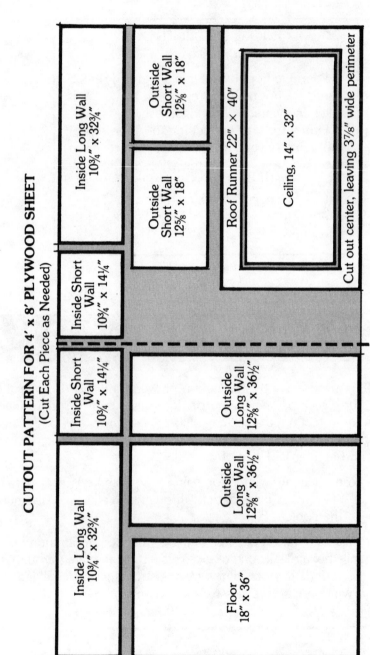

Inside Long Wall
10¾" x 32¾"

Outside Short Wall
12⅝" x 18"

Outside Short Wall
12⅝" x 18"

Roof Runner 22" × 40"

Ceiling, 14" x 32"

Inside Short Wall
10¾" x 14¼"

Inside Short Wall
10¾" x 14¼"

Outside Long Wall
12⅝" x 36½"

Inside Long Wall
10¾" x 32¾"

Outside Long Wall
12⅝" x 36½"

Floor
18" x 36"

Cut out center, leaving 3⅞" wide perimeter

Note: Sheet may be cut into two 4' x 4' sections
at the lumber yard for easier transporting.

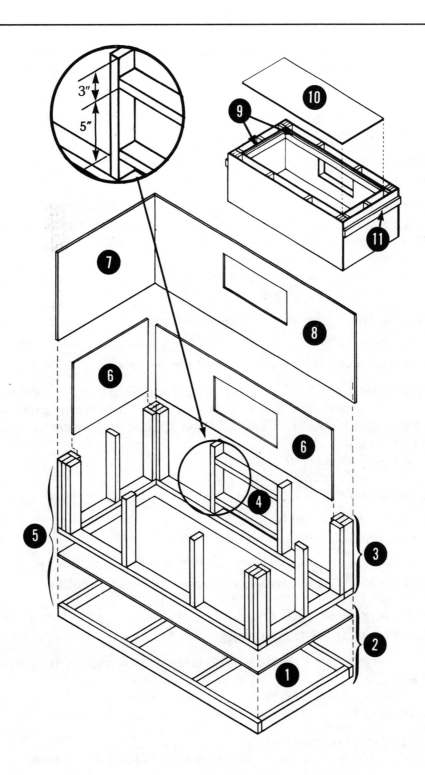

In this experiment you will build and weatherize what is practically a model of a real, one-story house. In Experiment No. 5, we will convert that house into a small solar garden. These experiments will illustrate some important energy principles.

As you can see, the experiments will require a considerable amount of wood and effort (and probably some occasional advice). Therefore, you may want to consider teaming up with several friends agreeable to sharing the work and the expense. The model house and solar garden would make an excellent class project, especially if you could use the school workshop.

Okay, let's get started. The numbered instructions below pertain to the numbered items on the drawing. Read them carefully. Good luck and happy hammering.

1. Make the Foundation.
   4 pieces of furring 16½" long
   2 pieces of furring 36" long

2. Place Floor on Foundation (don't nail it down yet).
   1 piece of plywood 18 by 36"

3. Construct Wall Frames by Erecting Studs on Sole Plates (nail corner studs together first, then anchor studs by nailing upward through sole plates; allow 11¼" between window studs).
   2 pieces of furring 14¾" long
   2 pieces of furring 36" long (the sole plates)
   22 pieces of furring 10" long

4. Nail Window Frames Between Window Studs.
   2 pieces of furring 11¼" long

5. Nail Wall-Frame Assembly and Floor to Foundation.

6. Hold an Inside Wall Against Window Studs and Trace Window Opening with a Pencil. Saw out Window Opening. Stretch Plastic Over Opening on Both Sides of Wall and Secure with Tape and Thumbtacks. Nail Inside Walls to Studs (install long walls first).
   2 pieces of plywood 10¾" by 32¾"
   2 pieces of plywood 10¾" by 14¼"
   2 pieces of plastic sheet 7" by 13"

7. Nail Outside Short Walls to Studs.
   2 pieces of plywood 12⅝" by 18"

8. Make Window Opening and Cover With Plastic as in Step 6. Before Nailing Walls to Studs, Fill Cavity Below Window with Loose Insulation.
   2 pieces of plywood 12⅝" by 36½"
   2 pieces of plastic sheet 7" by 13"
   Loose insulation (enough to fill)

9. Nail Ceiling Supports on Inside Walls (align bottom edges with top of window opening).
   2 pieces of furring 32¼" long
   2 pieces of furring 12¾" long

10. Place Ceiling on Supports.
    1 piece plywood 14" by 32" (cut from center of 22" by 40" panel)

11. Nail Handle Grips on End Walls.
    2 pieces of furring 18½" long

And that completes the basic structure. At this point you're just about ready to weatherize the house. But before you do, you'll have to check the thermal efficiency of the model as it now stands. So...

> ### QUICKIE EXPERIMENT "D"
> Testing the Unweatherized House
>
> This quickie experiment won't be all that quick. But it *will* be easy. Put six ice cubes in a saucer and set the saucer on the floor inside the model house.
>
> Replace the ceiling, noting the time on the clock. Periodically look through the plastic window to check the ice. When it has all melted, jot down how long the melting required.

Now you can weatherize the house. Cut the foam pads to fit the top of the ceiling and the spaces under the floor. Whatever remains can be used as a second layer under the floor, if space permits.

Secure the pads in place with glue and tape. Be sure to put the pads on the unfinished side of the plywood ceiling. Then fill the wall spaces with loose insulation, but don't pack it down. Run weatherstripping around the top of the ceiling supports. And finally, if the walls and floor don't fit together too well, get a small tube of caulking compound and caulk all inside joints.

How can you tell if the weatherized house is any more effective than it was before? That's right, run another experiment. Here we go again…

> ### QUICKIE EXPERIMENT "E"
> Testing the Weatherized House
>
> Pick out six ice cubes of the same size you used for Quickie Experiment "D", and follow the same procedure as in that test. This time, however, you'll be putting the saucer of cubes into a weatherized house. Once again, note how long it took for the ice to melt.

If the total amount of ice was the same in both experiments, you should have recorded a longer melting time with the insulated house. Did you?

SOLAR ENERGY

So far you have discovered some of the factors that affect your family's energy usage and you have seen how reducing electrical usage and weatherizing your home can help to conserve energy.

However, no matter how much we conserve, the fact remains that two of our most important sources of energy—oil and natural gas—are in short supply. Sooner or later (and many experts say it had better be sooner) we will have to find new sources of energy for the years ahead.

Many alternative energy sources are being considered. One of the most attractive is the sun—and no wonder—the sun pours more clean energy on the earth in two weeks than is contained in all the world's coal, oil and natural-gas reserves combined. And best of all, it's free.

In the next experiment we will convert our weatherized house into a solar garden that you can use to actually grow small plants (potted herbs, for example) outside during the winter. This working model, by the way, is based on a full-sized solar garden that did indeed produce crops in commercial quantities all year round in Michigan.

EXPERIMENT 5: Converting to a Solar Garden

THINGS NEEDED: The weatherized house and leftover materials from Experiment No. 4; eight 2¼" nails; aluminum foil; a small screw eye; six 2-pound coffee cans; flat black paint; six to eight 3" or 4" pots with various herbs or other small plants. Paint brush. Saw.

In the winter, a solar garden needs something in which it can collect and store the sun's heat during the day for release during the night. Our "something" will be water...water in coffee cans. The cans will help the water to absorb heat better if they are dark and nonreflective. That's the reason for the flat black paint. So begin by painting the cans inside and out.

While they're drying, you can put up the roof. For that job, the instruction numbers below correspond to the numbers on the drawing.

1. Nail Roof Base Together.
   4 pieces of furring 18 ¾" long

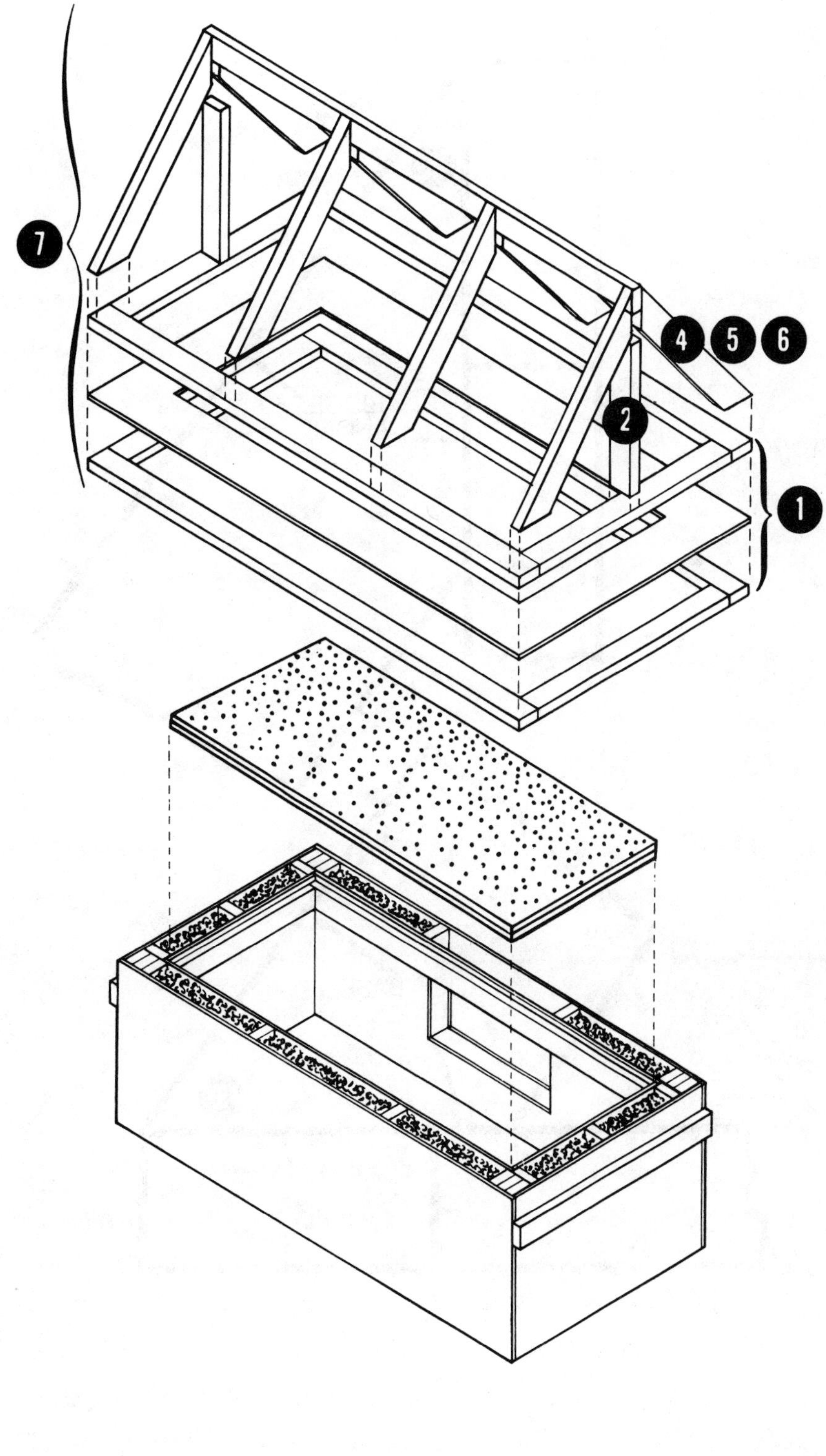
7
4 5 6
2
1

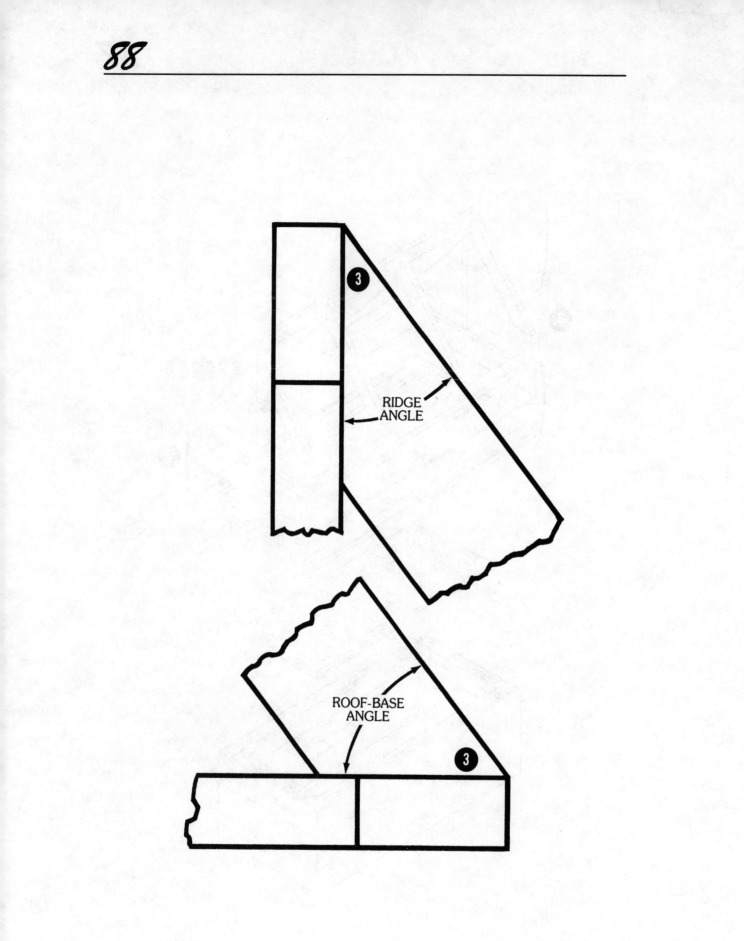

RIDGE
ANGLE

ROOF-BASE
ANGLE

4 pieces of furring 40" long
1 piece of plywood 22" by 40" (interior cutout dimensions: 14 ¼" by 32 ¼")

2. Erect Ridge Supports on Roof Base (anchor by nailing upward through the roof base; make sure that supports are centered).
2 pieces of furring 12" long
4 nails 2 ¼" long

3. Cut Out Cardboard Template for Marking Ridge and Roof-base Angles on Rafters.

4. Saw Rafters to Length.
8 pieces of furring 17 ¼" long

5. Mark Ridge and Roof-base Angles on Rafters. Saw Angles Off.
Cardboard template from Step 3
8 rafters from Step 4

6. Nail Rafters to Ridge Board.
1 piece of furring 40" long
8 angled rafters from Step 5

7. Nail Roof Assembly Together.
Roof base with ridge supports
Ridge board and rafters
4 nails 2 ¼" long (to fasten ridge to its supports)

That's it for the roof. Just a few more details and you can wrap up this project. But first let's check your handiwork. Get the house you built in Experiment No. 4 (whaddya mean, you can't find it?). If you worked with reasonable care, the roof overhang should just clear the exterior house walls. The plywood roof runner should cover the tops of all the walls and the same space in between. They do? Way to go.

Now for the final steps:

Glue aluminum foil to the entire underside of the ceiling. Try not to wrinkle the foil, as it must reflect the sun's rays into the garden when the ceiling is propped open.

Then set the ceiling and roof in place. Swing the window side of the ceiling upward until it hits the back rafters (you will have to raise the roof slightly to clear the front rafters). Next cut a length of 1" by 2" furring to keep it propped open to its maximum. A small nail in the ceiling will keep the prop from slipping.

Tightly cover the roof with the plastic drop cloth; if there are any wood projections that could tear the plastic, saw them off. Secure the plastic to the underside of the roof with tape and thumbtacks. You can lift the roof easily by grabbing the overhang with both hands, or you can insert a screw eye in the center of the ridge board. If you prefer the latter option, put a piece of tape over the plastic before twisting the screw in.

When starting up your solar garden, try to place it high enough off the ground so that it will not be covered by snow. A picnic table or bench would be great. Position the house so that the window faces south at high noon. Inside, arrange the coffee cans and potted plants as shown below.

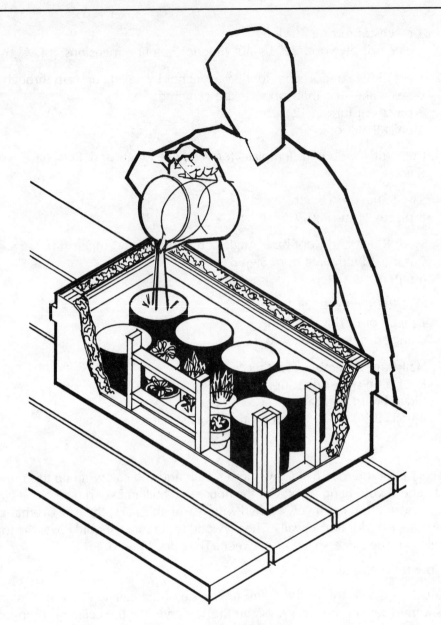

Fill the coffee cans with water.* You might also want to put a thermometer among the plants to monitor the temperature in your garden.

The ceiling should be propped open every day, even when the skies are cloudy. And it should be closed every night. In each case, the roof will have to be removed to adjust the ceiling, of course.

Your solar garden should provide a suitable temperature for the plants. But when the weather turns really cold, you may have to bring the plants indoors. Let the water in the coffee cans be your guide. If it starts to freeze, that's the time to move inside.

By the way, don't forget to maintain the water level in the cans; and remember to water those plants.

*Why use water? The best way to answer that question is to (oh-oh, here it comes) do another quickie experiment. Sooo...

QUICKIE EXPERIMENT "F"
Water as a Heat Storage Material

Try to get three small cans of the same size, like soup or tomato paste cans. Put water in one, sand in another, and pebbles in the last. Fill each to the same level.

If you have permission to do so, place the cans in an oven (*not* a microwave oven). Turn the thermostat to the lowest temperature possible and set the oven on "bake." After about a half hour, the contents of all cans should be at the same temperature. Using a pot holder, remove the cans and put them on some newspapers.

Wait several minutes. Then carefully touch each can. Keep doing that every so often. Which one holds heat the longest? Does that answer the question about using water in the coffee cans?

Thomas Edison, shown in his iron ore separating facility. An idea that proved uneconomical, and therefore unsuccessful at the time. Now, widely in use.

PART SIX

ALTERNATIVE ENERGY SOURCES

WHERE WILL TOMORROW'S ENERGY COME FROM?

Many concerned people are asking the above question today...and with good reason. As you learned earlier, two of our most important sources of energy—oil and natural gas—are in short supply. We must find new sources of energy that can serve our needs in the years ahead.

The search for alternative energy sources is a worldwide effort. Scientists working in industry, at universities and in government research laboratories are trying to unlock new supplies of energy. For although our planet has abundant reserves of energy, enough to last for countless years, much of this untapped energy is locked away beyond our present reach. We don't know how to get at it.

However, we are learning more about alternative energy sources every day. In the not-too-distant future, scientists will probably discover practical and economical ways to employ these different forms of energy.

The eight experiments that follow will introduce you to several of the more promising alternative energy sources. A few of these sources include sunlight, wind and geothermal energy. At the beginning of each experiment, you will find a brief introductory section that discusses the promises and problems of the particular energy source. Be sure to read it before you begin the experiment.

Good luck!

SOLAR ENERGY

Scientists estimate that the sunlight falling on the United States during a single summer day contains twice as much energy as our nation uses in an entire year! The problem lies in discovering how to collect this "free" energy in an efficient, economical manner.

The two experiments that follow demonstrate two direct methods of capturing solar energy. In other words, these methods transform sunlight itself into usable forms of energy: heat (Experiment No. 1) and electricity (Experiment No. 2).

However, energy from the sun can also be collected indirectly. For example, "windmills" (Experiment No. 3) and ocean thermal-energy systems (Experiment No. 4) convert the effects of sunlight into energy. The effects are wind and warm water. Sunlight provides the heat that makes the wind blow and warms the top layers of ocean water.

Solar energy has many advantages:

• It is an inexhaustible source of energy that will last as long as the sun itself.

• It is available in our own country. We do not have to import sunlight.

• It is a clean source of energy.

Solar energy also has serious disadvantages:

• Solar energy "turns off" at night and during cloudy weather.

• Solar energy is at its weakest during the winter months, the time of the year when homes and factories need energy the most.

• Using today's technology, solar energy is quite expensive.

However, the potential benefits of solar energy outweigh its shortcomings. Major research programs are underway all over the world to find ways to make solar energy practical.

EXPERIMENT 1: A Model Solar Hot-water Heater

THINGS NEEDED: About 10' of flexible black tubing. A shallow cardboard or wood box about 12" by 18". Flat black paint, dark paper or dark cloth for the inside of the box. A piece of window glass to cover the box. A spring-type wooden clothespin. Tape. Two empty cans or buckets.

This experiment will help you to understand how solar hot-water heaters operate. Many people are installing such systems in their homes today. Perhaps someone in your neighborhood has already done so.

The heart of a solar hot-water heater is a device called a "collector." It collects, or captures, solar energy and uses that energy to heat water. In a real solar hot-water heater installation, the collector is large and is mounted on the roof of the house. It must be aimed toward the south so that it collects the maximum amount of sunlight during each day.

You can easily build a model solar-energy collector. Start by covering the insides of the box with black paint, paper, cloth or other dark material. Then loop the tubing back and forth inside the box. Arrange for each end of the tubing to stick out of the box for 2'. Next, place the glass cover on the box and secure it with tape. We're now ready for a test.

Wait for a sunny day and find an open area. Place the collector on a stand, with a can or bucket of water next to the collector. Put one of the tubing ends into the water.

Suck gently on the other end to establish a siphoning action. Once water starts to flow through the tubing, pinch the tubing partially closed with the clothespin. This is done to limit the flow of water to a small trickle.

Notice that after a while, the water trickling from the tubing will be warmer than the water siphoning into the collector. Why? You're right. Solar energy absorbed by the tubing is transferred to the water, thereby heating it.

How warm the water gets will depend mostly on the time of year, how clear the skies are and how slowly the water flows through the collector.

EXPERIMENT 2: Electricity Directly from Sunlight

THINGS NEEDED: Silicon solar cell (see text). Block of wood. Compass. Cardboard. Small spool of magnet wire (#18 wire or finer). Two small alligator clamps. Soldering iron and solder. Tape. Glue. Soft old towel.

A silicon solar cell transforms light directly into electricity (in a manner too complicated to explain here). Many space satellites use solar cells for powering their scientific equipment.

Someday solar cells may be used to produce electricity for homes and factories on earth. This could happen if scientists figure out a way to manufacture large solar cells at a reasonable price. Right now the cost is too high.

However, you can use a small, inexpensive solar cell to demonstrate the process of transforming sunlight into electricity.

You can purchase a silicon solar cell for about $2 at most electronics stores. Treat it gently. Repeat: *Treat it gently*! It is extremely fragile and easily broken.

Since the cell comes without any connecting wires, you'll have to make a pair. Do this with 2 lengths of the magnet wire. Each connecting wire should be about 12" long. Carefully scrape ½" of the enamel insulation from both ends of both wires.

Now for the tricky part: Lay the solar cell on something soft, like an old towel. Solder one of the wires to the silver edge on the front of the cell. Do this very carefully. Then solder the other wire to the silver surface anywhere on the back of the cell.

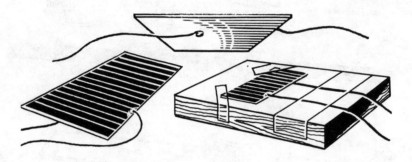

Finally, tape the cell and wires to your block of wood so that you can handle it conveniently.

Now set the cell aside. It's time to build a simple but very sensitive device—a galvanometer—to show that the cell can actually produce electricity. Making a galvanometer is easy. Turn back to page 6 for complete instructions.

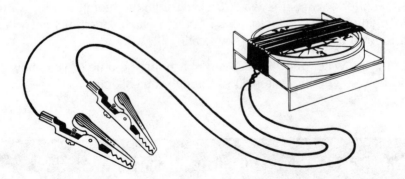

After you've installed the alligator clamps (see picture), connect the silicon solar cell and the galvanometer as shown. Arrange the compass so that the coil lines up with the needle.

When you expose the solar cell to light (either sunlight or the light from a flashlight or table lamp), the compass needle will move. Here's why: The electric current produced by the solar cell flows through the coil and produces a weak magnetic field in the E-W direction. This magnetic field tries to pull the compass needle in that direction.

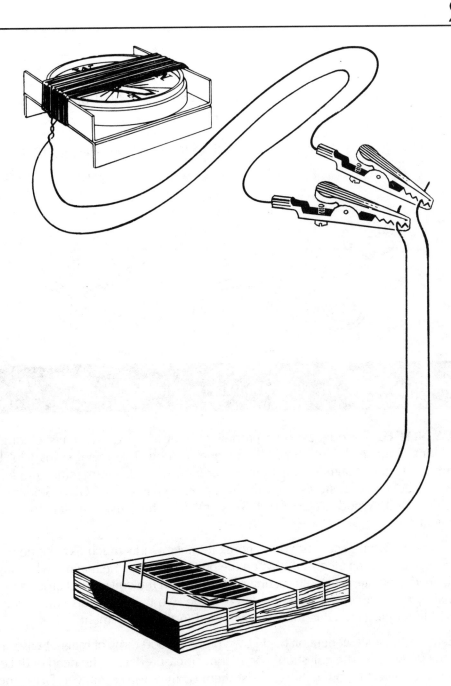

WIND ENERGY

Wind energy is really an offshoot of solar energy. Winds occur when different parts of the earth receive different amounts of heat energy from the sun.

Today many scientists feel that "windmills" designed to generate electricity can help meet future energy needs. Some say that homes will eventually have their own windmill generators.

Actually, the idea of harnessing wind energy is quite old. Wind was used many centuries ago to propel ships and to grind grain. American farmers pumped water with windmills in the 1920s and '30s.

Although wind energy is clean and appears to be free, there are many problems to be solved:

• The wind-powered generators available today are expensive.

- They are large.

- They do not generate electricity unless the wind is blowing, obviously.

- It is costly to store excess energy—that is, energy generated but not immediately needed.

For these reasons, experts believe that wind generators will be useful only in limited situations.

EXPERIMENT 3: Converting Wind Energy into Electricity

THINGS NEEDED: Model-airplane propeller about 6″ long. Two nails 1″ long. Two nails 3″ long. Four small nails. Small bar magnet 1″ long. Two metal strips 1 ½″ by 4″, cut from a tin can. Magnet wire from Experiment No. 2. Germanium diode, type 1N34A (most electronics stores carry this inexpensive part). Tape. Glue. Soldering iron and solder. Drill. Wood block about 3 ½″ by 5″. The galvanometer you used in Experiment No. 2.

The model wind generator you're going to make works much like the new "wind turbines" for generating electricity. When the propeller spins, the magnet whizzing past the nail head generates a tiny alternating current (AC) in the coil wound around the nail. The small germanium diode connected across the two nail terminals converts the AC into DC (direct current), which is what we need in this experiment.

To make the wind generator, begin by wrapping 1000 turns of magnet wire around one of the large nails. The coil should be 2″ long, measured from the head end. Leave a few inches of wire for connections. Twist them so they won't unravel. Drive this nail into the center of the wood block. Also drive in the two smaller nails where shown.

After scraping the enamel insulation off the ends of the coil wires, wrap the bared ends around the nail heads. Then hook the diode across the nails and make all connections secure by soldering.

Next, glue the bar magnet to the head of the other large nail. Be sure that the magnet is centered on the head and that the glue is given plenty of time to set. This will be our propeller shaft.

The shaft is supported by the 2 tin-can strips. Fold them in half lengthwise for added stiffness. Then bend them about ¾″ for the base. Nail them to the wood block, in line with the upright nail.

You'll have to decide how high the shaft holes should be. Locate the holes so that the magnet ends are close to the upright nail head but do not prevent the shaft from spinning freely.

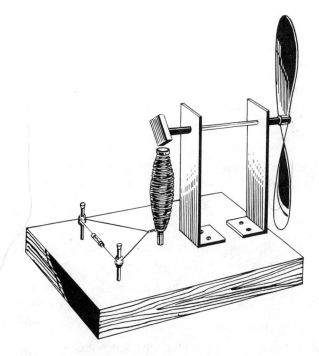

Insert the shaft in the supports until the magnet is directly over the nail head. Two collars of electrical tape will keep the shaft in place. Finally, drill a hole in the propeller so that it fits snugly on the nail.

Now for the test. Connect the galvanometer's alligator clamps to the 2 nail terminals. Keep the compass about 1' from the magnet. Again, as in Experiment No. 2, have the galvanometer coil lined up with the compass needle.

Set the generator in the wind or in front of a fan. (Or wrap a couple of feet of string around the shaft and pull upward, like when you try to start a power lawn mower.) When the shaft turns, you'll see the compass needle deflect. This demonstrates that electricity can be produced from the wind.

OCEAN THERMAL-ENERGY CONVERSION

Sunlight warms the top layer of water in the ocean. In some parts of the world, surface water temperatures are 80°F or higher.

However, the lower layers of water, hundreds of feet below the surface, are untouched by the sun's rays. Water at such depths usually has a temperature of about 40°F.

This difference in temperature can be used to drive a turbine-generator power plant. Here's how the process works, in simplified form: Let's say that the power plant uses ammonia as the working medium (this is not the same as household "ammonia," which is really ammonium hydroxide). The process begins with the ammonia in liquid form, at a temperature of 40°F.

1.   Warm seawater at the surface heats the liquid ammonia to a temperature of 80°F. This heating turns the ammonia into an expanding gas.

2.   The expanding gas rushes through the turbine, making it spin. The turbine is connected to an electric generator. As the turbine spins, so does the generator, thus producing electricity.

3.   Cool seawater, from the deep, chills the ammonia gas and turns it back into a liquid.

4.   Same as in Step 1...the cycle starts all over again.

Although this process could generate much electricity, building a large ocean thermal-energy conversion plant would be very difficult and expensive. Still, the concept holds promise in the warm parts of the world.

EXPERIMENT 4: The Idea Behind Ocean Thermal-energy Conversion

THINGS NEEDED: An empty narrow-mouthed plastic bottle. A balloon. A bucket filled with hot water. Another bucket filled with cold water.

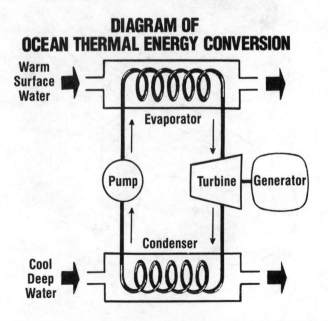

DIAGRAM OF OCEAN THERMAL ENERGY CONVERSION

This simple experiment uses air as the working medium to demonstrate the principle on which ocean thermal-energy conversion is based. Here's what you do.

Put the empty bottle and balloon in your refrigerator. Leave them there for at least one hour. They must be cold.

Before removing the bottle and balloon from the refrigerator, prepare the buckets of hot and cold water. Add a handful of ice cubes to the cold water to make it even colder.

Then take the balloon and squeeze it to drive out any air it may contain. After that, quickly slip the neck of the balloon over the mouth of the cold bottle.

Immerse the bottle in the hot water (it's okay if the balloon is in the water too). In a few moments you'll see the limp balloon begin to inflate.

When the inflation stops, immerse the bottle in the cold water. The balloon will slowly shrink and become limp again.

Repeat the hot and cold immersion steps.

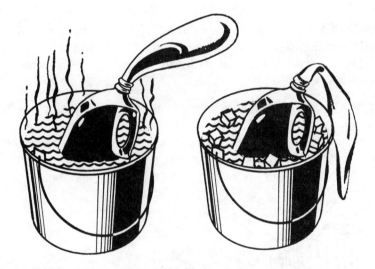

This experiment shows that heated water expands a working medium (the air inside the bottle) and makes it do "work" (blow up the balloon slightly). Cool water restores the medium to its original volume so that it can work again when reheated.

This experiment also demonstrates the principle of operation of a real ocean thermal-energy conversion system: The heated liquid turns into an expanding gas, which "works" by turning the turbine. When cooled, the gas condenses back into a liquid, shrinking in volume.

TIDAL ENERGY

In some parts of the world it's possible to harness the ocean's tides in order to make electricity. The idea is simple: First, build a dam across a bay. Then use the water that "piles up" against the dam to turn electric generators.

During high tide, the water builds up on the ocean side of the dam. It is allowed to flow into the bay by passing through turbines that turn electric generators.

During low tide, things are reversed. Now the water trapped on the land side of the dam again passes through the turbines as it flows back to the ocean.

Two tidal energy-conversion plants have been built overseas. They prove that the idea works. But not all bays are suitable for this kind of energy-conversion process. In fact, there are only about a dozen places in the United States where this kind of power plant could be built. The tides must be high enough to create a worthwhile difference in water height, and the bay must be narrow enough for a practical dam to be built.

ENERGY FROM TRASH

Turn trash into energy? Yes, indeed. Many cities across the United States are doing just that. The idea makes good sense.

After you perform Experiment No. 5, you will see that much of the waste we discard every day can be burned to produce heat. In turn, this heat can be used to generate electricity in a power plant.

Flammable materials are only part of the story. We also discard organic wastes (such as food scraps) that can be transformed into methane gas, the chief component of natural gas. In this way, our garbage can help supplement America's dwindling natural-gas supply.

The people of America throw away about 94 million tons of solid waste each year. But if you add the solid wastes produced by farming, farm animals, and industries such as logging and paper-making, the total climbs to 850 million tons. There's a lot of energy going to waste...in waste.

As you might expect, converting waste products into energy is an expensive process, particularly when it is done on a large scale. However, waste conversion kills two birds with one stone: First, it provides us with needed energy; second, it helps us dispose of waste materials. For both of these reasons, many experts predict that waste conversion will become a very popular practice in the years ahead.

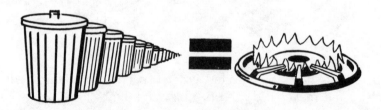

EXPERIMENT 5: Turning Trash into Usable Energy

THINGS NEEDED: Household trash (see text). A shallow baking dish. Aluminum foil. Scissors.

As we said earlier, many of the things we throw away every day can be burned to produce heat. They are a source of energy. This simple experiment proves that point. It shows one way of converting trash into fire fuel.

The first step is to put on a pair of gloves and rummage through your trash cans. Look for paper or cardboard items (that aren't too dirty or messy). For example:

• Can labels

• Toilet-tissue wrappers

- Cardboard boxes
- Paper cups or plates

- Paper towels
- Flour or sugar bags

You get the idea...the list can go on and on.

Using scissors, cut these items into pieces that will fit neatly into the baking dish. But don't put them into the dish yet. First soak them in warm water until they are soggy. While the paper pieces are soaking, line the dish with aluminum foil to keep it clean.

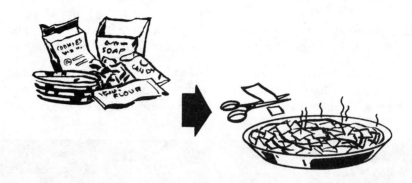

Then place layer after layer of soggy paper into the dish. Press the layers together and force the excess water out of the soggy mass. Pour this excess water out. Stop adding layers when you've built a pile that's about ¾" thick.

Now we want the compressed pile to dry out. For this demonstration only, let's speed up the drying process. Let's use an oven (better check to see if it's okay to use the oven for this purpose). Bake the pile for about an hour. The oven temperature should be around 200 degrees. *Don't use a microwave oven!* Because of the aluminum foil, the microwave tube inside the unit could be damaged.

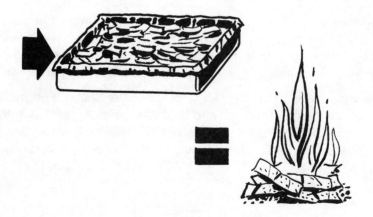

After taking the dish out of the oven and after the contents have cooled, lift the pile out of the dish. If it is still damp, set it aside until completely dry. When dry, chunks of this salvaged wastepaper will burn like wood. You can use them in a fireplace, camp fire or whatever.

When making additional piles, skip the oven part. (You don't need the baking dish either; use something else.) Simply let the piles dry outside in the aluminum-foil liner. It doesn't make sense to consume more energy in using the oven than you get from the fire fuel.

COAL CONVERSION

Coal is known as a fossil fuel because it is composed of the remains of trees and other plants that lived hundreds of millions of years ago. Oil and natural gas are also fossil fuels, all are available in America. But coal is, by far, the most plentiful of the three.

Coal has long been used as a fuel for electric power plants. Many years ago, coal was a popular fuel for heating homes. However coal is not as convenient to use as are oil and natural gas. The furnaces needed to burn coal are more complicated. And with coal, special equipment is needed to control air pollution in industry. For these reasons, scientists are searching for practical ways to convert coal into liquid fuel and gas.

Liquid fuel made from coal could be used in place of heating oil. It could also be processed into gasoline for cars and kerosene for jet planes. Gas produced from coal could be mixed with natural gas for home-heating and industrial use.

EXPERIMENT 6: Getting Methane from Coal

THINGS NEEDED: Lump of soft (bituminous) coal about the size of a baseball. A 2-pound coffee can. Funnel. Small glass bottle such as spices come in. Hammer. Old rag. Paper towel.

Fuel gas has been produced from coal for decades. The process requires high temperatures because chemical bonds must be broken to form the gas. (Experiment No. 7 illustrates that process.)

Coal also contains a small amount of burnable gas trapped within its pores and cavities. This gas is not chemically locked in the coal and can be released rather easily. You can prove this for yourself by doing the following simple experiment.

First, wrap the coal lump in a sturdy rag and hammer it into a fine powder. With a wad of soft paper, plug the narrow end of the funnel. Empty the coal powder into the funnel. Invert the coffee can over it. Then, pressing the funnel against the bottom of the can, return the can to its original upright position. Remove the paper wad.

Now gently pour water into the can until the water is about an inch above the funnel. (If it's a plastic funnel, you'll have to hold it down to keep it from floating.) The water, of course, will enter the funnel. Next, take the small bottle and submerge it, open end up, in the water. After it fills completely, turn the bottle upside down, keeping it

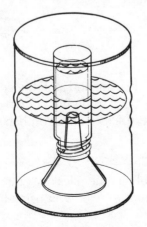

underwater so that no air gets into it. Then carefully maneuver it over the funnel—again, staying underwater. Finally, lower the bottle and let it rest on the funnel. That's all you have to do.

Put the can aside, and in a day or so you will see a large bubble at the upper part of the bottle. The bubble will contain mostly methane gas. It is the same kind of gas that formed when decaying plant matter began turning into coal eons ago, back when dinosaurs plodded the earth. Just think, that methane could have been trapped for millions of years...until you released it.

EXPERIMENT 7: Converting Coal to Fuel Gas

This experiment could be an optional classroom project, requiring your teacher's supervision.

THINGS NEEDED: Everything shown in the drawing.

If you are given permission to try this experiment in the school chemistry lab, set up the equipment as shown. Your teacher will guide you in the proper procedures.

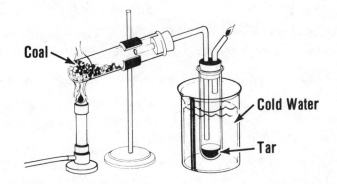

When all is ready, heat the coal in the test tube for several minutes to drive out the air. Afterward, bring a flame to the opening of the outlet tubing. The emerging gas will burn. This gas is known, not surprisingly, as coal gas. It consists of a number of burnable gases.

Along with these fuel gases, the heating of coal in the absence of air produces coke (the remains in the horizontal tube), coal tar (the remains in the upright tube) and ammonia. All are highly useful by-products of this process, which is still in use today.

GEOTHERMAL ENERGY

Scientists believe that the core of our planet is a large mass of molten material. They call this material "magma." It may have a temperature of 8000°F (whew!).

In most places on earth, the magma is many miles below the ground. But at some locations it comes close to the surface, where it creates hot spots. When ground water comes in contact with these hot spots, the water turns to steam. The geysers in Yellowstone National Park are well-known examples of this geothermal energy in action.

Incidentally, the name "geothermal" comes from geo (earth) and therm (heat). Geothermal means heat from the earth.

Where geothermal energy is available in the form of steam (at a suitable temperature and pressure), it is a practical source of energy. At the geysers' geothermal field in Northern California, enough electricity is generated to serve a city nearly the size of San Francisco.

The steam is used to spin turbines that drive electric generators. The process is not complicated, particularly if the steam is hot and dry. Moist steam, though, may carry minerals from the water. These minerals can clog and corrode the generating equipment.

The United States has 1.8 million acres of land where geothermal energy is known to exist in various forms. This leads to the belief that geothermal energy may eventually become an important source of electricity. Some estimates indicate that thirty percent of our electricity will come from geothermal energy by the twenty-first century.

EXPERIMENT 8: A Model Geothermal Steam Engine

THINGS NEEDED: An empty soup can. Heavy aluminum foil (frozen-food tray, pie pan, etc.). Aluminum wrapping foil. Straight pin. An 8" length of stiff wire (coat hanger). Stick about 12" long. Small pot. Glue. Tape. Rubber band. Scissors.

With a little imagination and a model turbine, you can get some idea of how steam can produce electricity. First you'll have to build the model. It's easy. Here's how:

1.   If not already off, remove the lid from the soup can. Throw the lid away; rinse out the can.

2.   Turn the can upside down and punch two ⅛" holes opposite one another in the bottom. Locate each hole about ¼" from the rim.

3.   From your frozen-food tray or pie pan, cut out a flat disk equal in diameter to that of the can. Pierce a hole in the center of this disk with a straight pin.

4.   Take some aluminum wrapping foil and wad it up into a little ball the size of a small cherry. Glue this ball to the disk, centered right over the hole. It will prevent the disk from wobbling badly.

5.   After the glue has dried, put the pin in the disk hole and push it through the ball. Get the pin as perpendicular to the disk as you can. Now enlarge this hole slightly with a thicker pin or needle. The idea is to have the disk and ball spin freely on the pin.

6.   Make 8 equally spaced "pie cuts" in the disk with a pair of scissors. Cut all the way to the ball.

7.   Twist each "pie wedge" slightly to form the turbine wheel (see drawing).

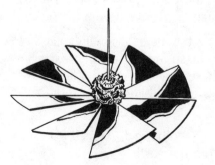

8.   Bend the stiff wire, as shown. Make sure that when the wire is later attached to the can, the small downward segment will point to the can's center.

9.   Slip the straight pin through the turbine and tape it to the support wire. Then tape the support wire to the can so that the turbine is as close as possible to the can without touching it. Blow on the turbine to test it. It should spin quite easily.

10.   Fasten the stick to the can with a rubber band or string.

11.   Now for the steam. Put a cup or two of water in a small pot. Before heating the water, cover the top of the pot with aluminum foil, pinching the edges all the way around. Make a pencil-sized hole in the center.

12.   Bring the water to a boil. When steam starts jetting from the hole, lower the can over the hole, using the stick as a handle. The steam pouring from the two holes in the soup can will start the turbine spinning merrily.

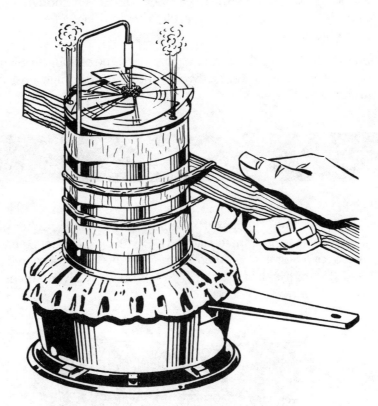

You'll have to imagine that in an actual power plant, high-pressure steam blasts against a series of such turbine wheels on the same shaft. The shaft, in turn, drives an electric generator, which produces the electricity. Isn't it amazing that steam can be so powerful?

THE FUEL CELL

What is a fuel cell? It is an electrochemical device for converting the chemical energy in fuels directly into electricity.

Why is it of high interest to scientists today? Because it can operate with two or three times the efficiency of other fuel-burning power producers. Thus the fuel cell, although not an energy source in itself, could help make our present fuel supply last years longer than expected.

Many scientists see the fuel cell playing a major role someday in providing electricity for automobiles and homes. Even now, large fuel-cell systems are being developed for use as power plants for shopping centers and similar sites. Small systems have already been used successfully in spacecraft.

In many ways, the fuel cell is like a battery. It contains no moving parts, is quiet, and wastes no energy as heat. Furthermore, it gives off no fumes. And, like a battery, it has electrodes and an electrolyte.

But unlike a battery, the fuel cell will continue to produce electricity as long as an oxidant, such as air or oxygen, can be fed to the electrodes. In most batteries, one of the electrodes serves as the fuel and is used up during operation, only the fuel cell's electrodes are never used up. Only the fuel.

There are several types of fuel cells. Some use hydrogen as fuel. Others use ammonia, alcohol or various hydrocarbon fuels. But even though they may work differently, all of them achieve the same result: They produce electricity directly from fuels in an efficient manner.

Unfortunately, however, scientists have to solve some serious problems before fuel cells can help ease our energy crisis. Two of these are the high cost and the short lifetime of the fuel cell.

EXPERIMENT 9: Making a "Fuel Cell"

THINGS NEEDED: A nonmetallic mesh tube, such as a woman's plastic hair roller, measuring about 3" long by 1" in diameter (see drawing). 3 small rubber bands or some string. Thin sheet of zinc or galvanized steel 3½" by 2". A large gauze pad (see text). A tablespoon or so of natural powdered graphite (one tube of powdered graphite lubricant should do the trick). Flour. Salt. 2 small bowls. Galvanometer from Experiment No. 2.

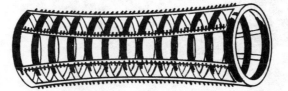

Making a fuel cell would be easy if we had the right materials. Trouble is, the right materials are either dangerous or difficult to get in small quantities. So let's take an easier path. Let's build a device that is considered to be a type of fuel cell. It uses zinc as the fuel (one electrode), air as the oxidant (the other electrode) and saltwater as the electrolyte.

First, pour all the graphite into a bowl. Next, in a separate bowl, prepare a binder of 1 part flour and 4 parts water. Then add the binder to the graphite, a little at a time, and stir. What we want to do is to end up with a thick paste, like peanut butter.

Now from the gauze pad cut two strips as wide as the mesh tube (hair roller) and about 7" long. If you don't have gauze, try a paper towel; fold it in 4 layers and cut it to size.

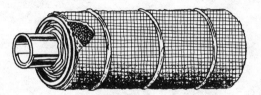

On a portion of one of these strips, spread all the graphite paste (messy, isn't it). You need cover only the first 3" of length, but cover the entire width of this 3" portion. Then place the mesh tube on the graphite and wrap the strip around the tube. The graphite will be facing the meshwork, of course.

Put the rubber bands or string around the gauze, and set the unit aside to dry (overnight, at least). That takes care of the air electrode for our "fuel cell." You may be wondering what part the graphite plays in the air electrode. Without getting too technical, it allows the oxygen in the air to enter the electrochemical reaction.

For the zinc electrode, roll the length of the zinc or galvanized steel into a tube small enough to fit inside the air electrode (but don't insert it yet). Then wrap the second gauze strip around the zinc.

After the air electrode has had a chance to dry out, insert the wrapped zinc electrode into the tube core. It should be a snug fit.

We still need an electrolyte. So dissolve a tablespoon of salt in two or three tablespoons of hot water, and pour this electrolyte into a saucer. That completes all the preparations for the fuel cell. Now, will it work? Let's see.

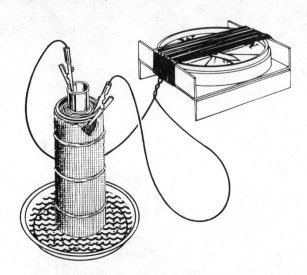

The inventive genius Edison with the scientific genius Charles P. Steinmetz at Schenectady, New York, October 18, 1922.

PART SEVEN

NUCLEAR EXPERIMENTS

THE DANCE OF ATOMS

Way back in 1922, before Einstein developed the theory of relativity, Thomas Edison said, "One day electricity will be produced by atomics." How, with his limited formal education and with so little of nuclear theory understood at that time, could he have made so accurate a prediction? We will never know.

While Edison did not perform any nuclear experiments of his own, it seems only natural to assume that he surely would have had he lived at a slightly later period in history. Certainly the readers of this book, which include a few Edisons of the future, should be conversant with the amazing properties and energy of the atom. Nature provides many examples for us to readily observe these properties. The sun, being a nuclear reactor, is one example. Radioactivity is another.

Most of the nuclear power produced in the world today comes from the controlled splitting (fission) of radioactive uranium. In the process, a reaction occurs that transfers uranium's nuclear energy into heat. Heat, in turn, creates the steam with which turbines generate electricity, propel ships and drive industrial processes.

As a fuel, uranium has a distinct advantage over coal, oil and gas. The latter three, as you know, are in limited supply. On the other hand, the known reserves of uranium would provide the world with energy for centuries to come. It represents a heat source that, properly controlled, is safe and does not significantly affect our environment.

This section of the book contains some of the basic facts about nuclear energy, along with experiments related to these facts. The eight experiments presented in this section will help you to understand a few of the fundamentals about the tremendous energy locked within the atom. Before you begin to experiment, though, you should know a bit more about atoms.

WHAT IS AN ATOM?

All matter is made of atoms...different kinds of atoms joined in different combinations. The page you are reading is made of zillions of atoms. So are you. And so is everything around you. An atom is an exceedingly tiny thing: 200 million atoms lying side by side would span a distance of only one inch.

As late as the nineteenth century, many leading scientists thought that atoms were indivisible blobs of matter, sort of like tiny, solid balls. Now we know that atoms are far more complex than that. We also know that under the right conditions, certain atoms can be split into smaller atoms.

An easy way to picture an atom is to think of it as a miniature solar system. In the center (somewhat like our sun) there is a relatively large structure called the nucleus. Whirling around the nucleus (somewhat like planets) there are tiny particles called electrons. Each electron carries a negative electric charge.

Atoms of one material differ from atoms of another material because of the make-up of their nuclei and the number of electrons each atom has.

The nucleus of most atoms consists of two kinds of particles: protons and neutrons. Both the proton and the neutron have about the same size and weight. However, the proton carries a positive electric charge, while the neutron has no charge at all.

Note that we said the nucleus of most atoms contains both protons and neutrons. The exception is the hydrogen atom, the simplest atom of all. Its nucleus contains but a single proton.

The number of electrons orbiting an atom's nucleus is equal to the number of protons in the nucleus. Because each electron carries a negative charge and each proton carries a positive charge, the charges balance. This makes the atom electrically neutral, which is another way of saying that the total negative charge equals the total positive charge.

RADIOACTIVITY

Interestingly, the neutron and proton were not discovered until well into the twentieth century. However, around the turn of this century, scientists observed that certain atoms undergo mysterious transformations. For example, atoms of radium (a rare metallic element) turn into atoms of radon (a rare gaseous element).

Equally as surprising, the radium atom emits a tiny particle, called an "alpha particle," when the change takes place. This alpha particle consists of two neutrons and two protons. It is identical to the nucleus of a helium atom. A stream of alpha particles is called an "alpha ray."

We now know that this transformation is an example of radioactive decay, a process whereby one atom breaks apart to form one or more smaller atoms.

Radium atoms decay into radon atoms and alpha particles spontaneously. In other words, every so often an atom in a chunk of radium metal will simply break apart. There are many other atoms that will break apart spontaneously. Scientists call these substances "naturally" radioactive.

When an atom breaks apart, the decay process also gives off energy. In certain circumstances it is possible to capture this energy in the form of heat, then use the heat to generate electricity. This is the principle of operation in the nuclear-powered electricity-generating plants that are at work today.

ATOMIC COUSINS

There's one more nuclear term we want to define here: "isotope." A bit earlier we said that different atoms have different nuclei and different numbers of electrons. An atom of oxygen, for example, has eight protons and eight neutrons in its nucleus and eight electrons whirling around the nucleus. Broadly speaking, it is the number of protons and electrons that determines the character of an atom, which in turn, determines the character of the chemical element it forms.

Sometimes a specific chemical element will contain atoms of slightly differing form. The number of protons and electrons will be the same, but the number of neutrons will not. These slightly different atoms are called isotopes. There are, for example, isotopes of oxygen that contain seven and nine neutrons instead of eight.

Isotopes play an important role in nuclear energy. Specifically, an isotope of uranium called U-235 was the first material used to create a nuclear chain reaction.

If you have followed what we've said so far, you are ready to learn more about nuclear energy by performing the experiments that follow. Good luck!

EXPERIMENT 1: An Oil-drop Model of a Splitting Atom

THINGS NEEDED: A small water glass. 5 or 6 ounces of rubbing alcohol. An ounce or so of cooking oil. Some water. A teaspoon and a butter knife. A paper towel.

Many scientists have suggested that a splitting atom behaves somewhat like a drop of liquid when it breaks up into droplets. This experiment demonstrates the point.

Fill the water glass about half full with rubbing alcohol, then add enough water to fill the glass two-thirds full. Stir the alcohol-water mixture with the teaspoon. Next, wipe the teaspoon dry and fill it with cooking oil.

Now comes the tricky part: Carefully bring the spoon close to the surface of the alcohol-water mixture in the glass, then gently tip the spoon over. If you've done the job right, a single blob of oil will slide into the glass.

If the blob of oil is floating on the surface, carefully add a bit more alcohol to the mixture (use your teaspoon); if the blob has sunk to the bottom of the glass, spoon in some more water. The idea is to change the blob of oil into an oil drop that hovers somewhere in the middle of the glass, as shown in the drawing.

Note how perfectly spherical the drop is. The forces that hold the oil drop together are analogous to the forces that hold an atom together.

Now take the butter knife and carefully prod the drop apart. At first the drop will bulge. Then it will tear apart into two perfectly round oil drops. The oil-drop "atom" will have split into two smaller "atoms." Note that the drop wouldn't split until it was critically deformed by the knife. Atoms behave in much the same way. They resist splitting until some action critically deforms them.

EXPERIMENT 2: A Domino Model of a Chain Reaction

THINGS NEEDED: A set of dominoes.

Earlier we talked about atoms that decay spontaneously into smaller atoms. Inside a nuclear power plant, though, atoms are made to split. And this splitting occurs more or less on schedule rather than by accident.

The isotope of uranium, called U-235, is ideally suited for such action. U-235 atoms are easily split by bombarding them with neutrons. In effect, the neutrons act like bullets that trigger the splitting of the uranium atoms.

When a neutron strikes a U-235 atom, several things happen:

1. The atom breaks into the smaller atoms of barium and krypton.

2. A substantial amount of energy is released.

3. Two or more neutrons are hurled away by the splitting atom.

Item 1 is not too important because we aren't really concerned here with the by-products of the split uranium atom.

Item 2 is very important. This energy will be converted into electricity (we'll see how in a later experiment).

Item 3 is absolutely vital, for these emitted neutrons make it possible to produce a steady stream of nuclear energy. How? When one U-235 atom splits, the neutrons it releases cause other U-235 atoms to split. The additional neutrons released trigger still other U-235 atoms. On and on it goes. This kind of process is called a chain reaction.

Imagine a chunk of U-235 in which a chain reaction has begun. If the reaction takes place quickly enough, an enormous amount of energy is released.

You can demonstrate this type of rapid-fire chain reaction by setting up your dominoes as shown in the first domino drawing. When you tip over the leading domino, as if shooting a neutron bullet into uranium, it tips over two other dominoes (releases two new neutrons). In turn, the two falling dominoes tip four more, and the uncontrolled chain reaction proceeds to completion.

However in a nuclear power plant, a runaway chain reaction—that is, an "atomic-bomb" explosion— is impossible. Nuclear power plants control the reaction. Here's how it's done:

The heart of a nuclear power plant is a nuclear reactor. Very simply, a reactor contains bundles of nuclear fuel (U-235) separated by materials that absorb neutrons. Thus, when a U-235 atom splits, all but one of the neutrons are absorbed before they can reach other U-235 atoms. The single remaining neutron is available to split another U-235 atom.

The result is a steady release of energy over a long period of time...a chain reaction that lasts for years instead of for a fraction of a second.

You can model this slow-moving kind of chain reaction by setting up your dominoes as shown in the second domino drawing. This type of chain reaction "wastes" some of the neutrons produced. Some dominoes fall without hitting other dominos.

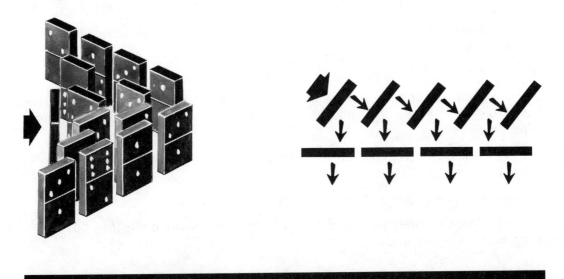

EXPERIMENT 3: Observing Radioactivity with an Electroscope

THINGS NEEDED: A source of alpha rays.* 2 water glasses (one about 1" taller than the other). An old 12" phonograph record. A piece of wool cloth. A 4" length of ¼" diameter wood dowel. A 6" length of clean stiff wire. Aluminum foil. Foil from a chewing-gum wrapper.

*The experiments herein require an alpha-ray source and a gamma-ray source. Suitable sources not exceeding the low radioactivity limits set by the U.S. Nuclear Regulatory Commission are available to the public.

Information on inexpensive "license-free" or "exempt-quantity" sources can be obtained from The Nucleus, Inc., Box 2561, Oak Ridge, TN 37830. This company provides an alpha-ray source containing polonium-210, designated as PO-210-S(alpha). It also provides a gamma-ray source containing cesium-137, designated as CS-137-S(Gamma). The mail-order cost is $15.00 apiece.

Each is a solid source housed within a 1"-diameter by ¼" thick plastic disk with an identifying label. S-6 is completely sealed. But since sealing would block alpha rays, S-2 is uncovered. Therefore the user is cautioned not to disturb the polonium-210 coating recessed within the disk.

These solid sources are regarded by the supplier as safe. But as with all chemicals and tools, radioactive materials should be respected and used with care. A brochure on proper handling techniques and safety precautions comes with the sources.

An even less expensive low-level source of alpha, beta and gamma radiation is the silk gas mantle used for ordinary camping lanterns. Readily available in stores selling camping equipment, gas mantles cost around 60 cents for a bag of two. We recommend they be left in their sealed plastic bag.

We were also told that coal, slate and granite are possible low-level radioactive sources.

Your school may have a gamma ray source or a chunk of radioactive ore. If not, try to find an old luminous-dial watch or clock. These timepieces, unlike newer models, have a small amount of radioactive material, radium, mixed in with the dial phosphors. If all else fails, a low level gamma ray source intended for educational use is available by mail order.

This experiment is similar to a late nineteenth-century method of observing the effects of radioactivity. It's based on the fact that alpha particles can discharge an object that has been charged with static electricity. The principle isn't hard to understand.

When alpha particles strike atoms in the air, they "peel" electrons away from the atoms, leaving positively charged ions. These ions (and the negatively charged electrons) are then available to discharge the static electricity on any nearby object.

In this experiment, a simple electroscope indicates the presence of static electricity on the surface of a phonograph record. Make the electroscope as follows:

1. Twist one end of the wire around the center of the dowel. Leave about 3" of wire protruding.

2. Bend a small L-shaped hook in the end of the wire.

3. Wrap and crumple some aluminum foil around the middle of the dowel. The ball of foil you make should be about ¾" in diameter; the foil must make good contact with the wire.

4. Soak the chewing-gum wrapper in warm water to remove the paper liner. Cut a small strip about 2" by ⅜".

5. Fold the strip in half, then place it over the hook.

6. Carefully insert the entire assembly into the smaller glass; the strip should hang freely in the center.

Next, put the taller glass about 4" from the electroscope. Rub one surface of the record with the wool cloth for several seconds to charge it with static electricity. Then place the record, charged side down, on the taller glass. The rim of the record must hang over the foil ball on the electroscope.

As soon as you bring the charged record into postion, you will see the 2 foil leaves of the electroscope spring apart. They are reacting to the record's static electric charge.

Now bring the alpha radiation source close to the electroscope ball under the record, with the opening of the source pointing at the ball. As the alpha particles discharge the static electricity (by producing ions in the surrounding air), the foil leaves will drop to their original position.

If you rotate the large glass slightly, you will shift a still-charged part of the record's surface over the electroscope. The leaves will fly apart again.

Repeat the experiment several times, holding the alpha source at different distances each time. You will observe that the farther the source is from the record, the longer it takes to discharge the static electricity over the foil ball. (Note: Before dismantling your equipment, read Experiment No. 7.)

EXPERIMENT 4: Observing Radioactivity by Radiography

THINGS NEEDED: Alpha-ray source from Experiment No. 3. Photographic print paper and a "one-shot" kit containing developer, stop-bath and fixer (both paper and kit are available at photo-supply shops). A 16-ounce measuring cup. 3 small plastic trays. A photographic "safelight" (optional). A darkroom to work in (a bathroom is ideal). Plastic tongs.

This is an experiment in radiography, which is the art of taking a picture with radiation other than light—X rays, for example. In this case, we'll use alpha rays. Here, photographic print paper exposed to alpha rays will, when developed, show the radiation as...well, you'll see.

If you do not have access to photographic lab equipment and decide to buy the materials needed, start by mixing the three chemical powders supplied in the kit. Follow directions, but don't worry too much about water temperature; the quality of the developed print won't matter. Wash your hands if you get any powder or solution on them; some people are sensitive to the chemicals and can develop minor skin rashes if exposed to them.

Line the 3 plastic trays in a row. Put developer solution in the first tray, stop-bath solution in the second tray, and fixer in the third tray.

If you have a safelight, you can use it to illuminate the room as you perform the rest of the experiment. If not, do everything in total darkness.

1.   Remove a sheet of print paper from the package. Reseal the package to shield the rest of the paper from light.

2.   Place the alpha-ray source on the paper; be sure that the opening is pointing down. The rays must strike the emulsion side of the paper (the creamy, smooth side). Leave the source in place for at least one minute.

3.   Set the alpha source aside, then slip the paper into the first tray (the developer solution). Leave the paper in the solution for about 90 seconds. Use the tongs to jiggle the paper every 15 seconds or so.

4.   Pick up the piece of paper with your tongs and transfer it to the second tray (the stop-bath solution). Leave the paper in that tray for 20 seconds.

5.   Now, with your tongs transfer the paper to the third tray (the fixer solution). Leave it in the solution for at least 5 minutes. You can turn on the room lights after the paper has been in the third tray for 30 seconds.

6.   Rinse the completed print under cold running water for at least 5 minutes.

If you've followed these instructions carefully, the print will show a darkened circular area corresponding to the opening in the alpha source.

The longer the source was left in contact with the paper, the darker the area will be.

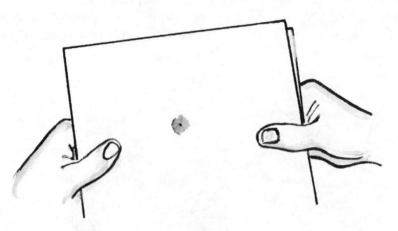

Note: Before disposing of the photographic solutions, read Experiment No. 7. The solutions are good for at least 12 print runs. But before each run, be sure to rinse off all traces of fixer solution on the tongs.

EXPERIMENT 5: Observing Radioactivity with a Cloud Chamber

THINGS NEEDED: A piece of dry ice (see paragraph 4 below). A large glass jar with a lid; the jar should be about 5″ high and about 4″ wide (an old peanut-butter jar would be fine). A piece of thick blotting paper. Glue. A piece of black-velvet cloth. An old towel. Some rubbing alcohol. A powerful flashlight or high-intensity desk lamp. Alpha-ray source from Experiment No. 3.

A cloud chamber is a device that allows scientists to see the trails made by nuclear particles. It was invented in 1912 by Charles T. R. Wilson, a pioneer atomic physicist.

Wilson discovered that water vapor can condense on ions just as it does on bits of dust to form raindrops. Now, since nuclear particles, such as alpha rays, produce ions as they streak through water vapor, the vapor that condenses on these ions shows up as fine, whitish trails.

The cloud chamber in this experiment is a modification of Wilson's original design. It is called a "diffusion-type" chamber, and it uses alcohol vapor instead of water vapor.

Look up "Dry Ice" in your Yellow Pages telephone directory to find a dealer in your area. You need a piece about 6" by 6" by 2" (roughly 2 pounds in weight). Dry ice is frozen carbon dioxide. It is cold enough to cause severe burns to unprotected skin and must be handled very carefully. Wrap the dry ice in a towel; never touch the exposed ice. For safety's sake, it is a good idea to wear leather gloves when you handle the block.

Make the cloud chamber as follows:

1. Cut a circle of black-velvet cloth to fit inside the metal jar lid. Cement it in place with a few dabs of glue.

2. Cut a circle of blotting paper to fit inside the bottom of the jar. Cement it in place with a few dabs of glue.

3. After the glue has dried fully, drip some rubbing alcohol onto the blotting paper. Keep dripping until the blotting paper is completely saturated but does not show an excess of alcohol on its surface.

4. Screw the lid on the jar. Then place the jar, lid side down, on top of the wrapped block of dry ice.

5. Darken the room completely, then position the flashlight as shown in the drawing.

6.   Wait patiently and keep looking at the path of the light beam.

After several minutes (it takes the chamber several minutes to form its cloud), you will see occasional white tracks near the metal lid. These are produced by the passage of cosmic rays (radiation from outer space) through the cloud chamber.

The alpha source used in earlier experiments is not an ideal source for use inside a cloud chamber. (While the "source" is an exempt quantity and can be discarded in the garbage when you are through with it, like all radioactive materials, chemicals, etc., they should be handled with care at all times.) However, it will work (usually) if you modify it as follows:

1.   Tape a piece of aluminum over the opening on the source.

2.   Using a pin or needle, pierce a small hole through the tape and foil. *Be very careful not to let the pin or needle touch the radioactive material at the bottom of the opening.*

Open the cloud chamber (it will be cold, so take care). Resaturate the blotting paper with rubbing alcohol. Place the alpha source on the black-velvet so that the hole points inward. Screw on the glass bottom, place the cloud chamber back on the dry ice, and wait. If all goes well, you will see fat white tracks characteristic of the passage of alpha particles.

EXPERIMENT 6: A Model Nuclear Power Plant Steam Turbine

THINGS NEEDED: A small unopened can of your favorite fruit juice. An empty soup can. A clean finishing nail. A wire coat hanger. A can of Sterno canned heat. An eyedropper. Two small sheet-metal screws. Hammer. Tin snips.

The core of a nuclear reactor (the part that contains the uranium) generates heat as the chain reaction takes place. Nuclear power plants boil water with this heat. They then use the resulting steam to drive turbines that, in turn, drive electric generators. In this way, the energy emitted by the splitting atoms is converted into electricity.

This simple model will show you how a steam turbine operates. Here's how to build it:

1.   Use the nail to punch 2 small holes in the top of the fruit-juice can. The holes should be on opposite sides of the top, about ¼" from the edge.

2.   Pour all of the fruit juice out of the can (maybe you'd like to take a juice break at this point). Remove the paper label, then rinse the can with water as best as you are able.

3.   Plug one of the 2 holes (either one) with a sheet-metal screw. And that's our water boiler.

4.   Now carefully remove the bottom of the soup can with a can opener. Save the bottom.

5.   Using a pair of tin snips, cautiously cut along the length of the can and flatten out the metal. Look out for those sharp edges.

6.   From the flattened metal, cut a strip about 4½" by ½". Again, work with care.

7.   With the finishing nail, hammer a hole about ¼" from each end of the strip and one in the very center.

8.   Carefully bend the strip into a square-cornered "U" bracket. The end holes in the strip must line up opposite each other. Make the bottom of the "U" about ½" wide.

9.   Take the lid you set aside earlier and punch a small hole in the exact center. The hole must be large enough that the lid spins freely on the finishing nail.

10.   Make 8 equally spaced "pie" cuts into the can lid, as shown in the drawing.

11.   Gently bend the cut sections to create an 8-bladed turbine wheel.

12.   Using the finishing nail as an axle, assemble the fan inside the "U" bracket.

13.   Locate the bracket on the can so that when steam shoots from the opening, it will hit the flat part of the blades. Now mount the bracket on the can with the other sheet-metal screw. To keep the turbine wheel centered, wrap some tape on the axle on both sides of the wheel.

14.   Fashion a simple support stand out of the coat hanger, as illustrated. The stand must support the juice can about 4" over the open Sterno can.

15.   Use the eyedropper to fill the can about ⅓ full with water.

16.   Light the Sterno and place it under the boiler. In a few minutes the water will boil and steam will spin the turbine.

EXPERIMENT 7: Demonstration of how Radioactivity can be Shielded

THINGS NEEDED: Equipment from Experiments Nos. 3 and 4 (see also the footnote on page 115.) Small pieces of different materials such as paper, aluminum foil, plastic and wood.

Certain radioactive emissions can be dangerous. Thus nuclear power plants use extensive shielding to protect their employees and the people living in nearby communities. This shielding is so effective that you can spend your whole life living next to a nuclear power plant and receive only about the same dose of radiation as you would if you had been given a chest x-ray.

This experiment demonstrates the radioactivity-shielding properties of various materials. Before starting you should know a little more about the three kinds of radiation given off by a radioactive substance: They are alpha rays, beta rays and gamma rays.

As we said earlier, alpha rays are made of alpha particles, and each alpha particle is identical, in all respects, to the nucleus of a helium atom. Because of its relatively large size (compared with other forms of radiation), an alpha particle is easily blocked by many common materials.

Similarly, beta rays are made of beta particles. A beta particle is actually an electron. It is much smaller than an alpha particle and therefore has greater penetrating power. Denser materials are needed to stop beta particles.

But of the three, gamma rays have the greatest penetrating power and they require the densest shielding materials. Curiously, gamma rays are not made of particles. There is no such thing as a gamma particle. Gamma rays consist of individual packets of energy called "photons."

Now to work. The two experiments you performed earlier, Experiments Nos. 3 and 4, are easily modified to demonstrate how alpha rays can be shielded. All you need do is to cover the opening on your alpha source with different materials before you perform the experiments. If a particular material blocks alpha rays, the static electricity around the electroscope (Experiment No. 3) will not be discharged and the photographic paper (Experiment No. 4) will show nothing.

Test as many different materials as you can find; keep a log that records whether or not each specific material blocks alpha rays. Try to include a thin sheet of mica in your test. Perhaps your chemistry or science teacher may be able to lend you a piece. You will find that this material allows alpha rays to pass through.

Testing materials for their ability to stop gamma rays is not as simple as the alpha-ray experiment. For this you need a gamma source and a Geiger counter, which detects radioactivity. If, as suggested in Experiment No. 8, your class has built a Geiger counter and has a gamma source, you can investigate a much wider range of materials. A few possibilities include concrete, steel, water and lead.

EXPERIMENT 8: Build a Geiger Counter (a Class Project)

THINGS NEEDED: Geiger-counter components (see parts list and circuit diagram). Gamma-ray source. Geiger tube (A tube and bakelite socket are available for $40.00 from The Nucleus, Inc. Box 2561, Oak Ridge TN 37830. Ask for tube-socket unit SG-2.)

Chances are you've heard of a Geiger counter before. This versatile instrument is one of the most useful pieces of equipment ever developed for detecting the presence of radioactive emissions.

A Geiger counter is a relatively simple device. Its heart is the Geiger tube, a chamber filled with a mixture of special gases and equipped with a pair of internal electrodes that must be charged at a high voltage.

Normally the gas inside a charged Geiger tube does not conduct electricity. However, if a radioactive emission strikes the tube, the gas is momentarily ionized and it becomes a conductor for a split second. Thus a pulse of electric current flows through the tube, from electrode to electrode. The electrical circuit of the Geiger counter is designed so that a click is produced by the headphones each time a pulse of current flows through the tube. Each click you hear means that a radioactive emission has struck the Geiger tube.

Some Geiger tubes will respond to alpha, beta and gamma radiation. However, the tube we used has a thin metal envelope that cannot be penetrated by alpha particles. Thus this instrument detects only beta and gamma rays, along with cosmic rays from outer space.

The Geiger tube needs about 900 volts in order to work. And believe it or not, our little unit can generate that voltage from the 6-volt battery. But don't let that high voltage frighten you. The unit is completely safe because it has a very low current output. Nevertheless, don't touch the Geiger tube or the capacitors. You can get a mild shock if you go looking for trouble.

How does the unit build 6 volts into 900 volts? In steps. Tapping the charge button (located on top of the unit) allows the voltage to climb with each tap. Eventually the voltage reaches operating level. More about this later.

Build the Geiger counter on a piece of perforated chassis board. Use push-in terminal strips to support the various components. The diagram shows how the parts must be connected. Keep these points in mind as you work:

1. The transistors will not operate, and may be damaged, if not connected exactly as shown. This is also true of the silicon rectifiers.

2. Try not to overheat the transistors and rectifiers when you solder them in place. It's a good idea to grip each lead with long-nosed pliers when you start to solder. (The pliers act as a "heat sink" to protect the parts.)

3. All solder joints must be bright and shiny. Gray, dull, grainy solder joints usually mean poor electrical connections and could prevent proper operation.

4. The Geiger tube is so delicate that you could accidentally deform it with your fingers, so mount it carefully. Use "broom clips," one at each end. Also, make sure you connect to the right leads on the Geiger-tube socket. The Geiger-tube pins are numbered on the underside of the base. Pin 1 connects to the main circuit and Pin 3 connects to ground. Pin 2 is a dummy pin that is not used.

5. Be certain that the negative terminal of the battery goes to ground.

GEIGER COUNTER CIRCUIT DIAGRAM

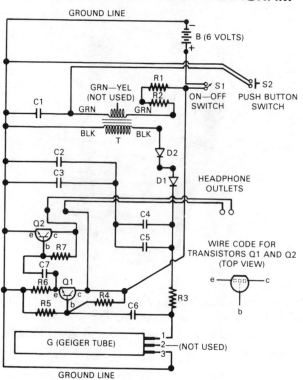

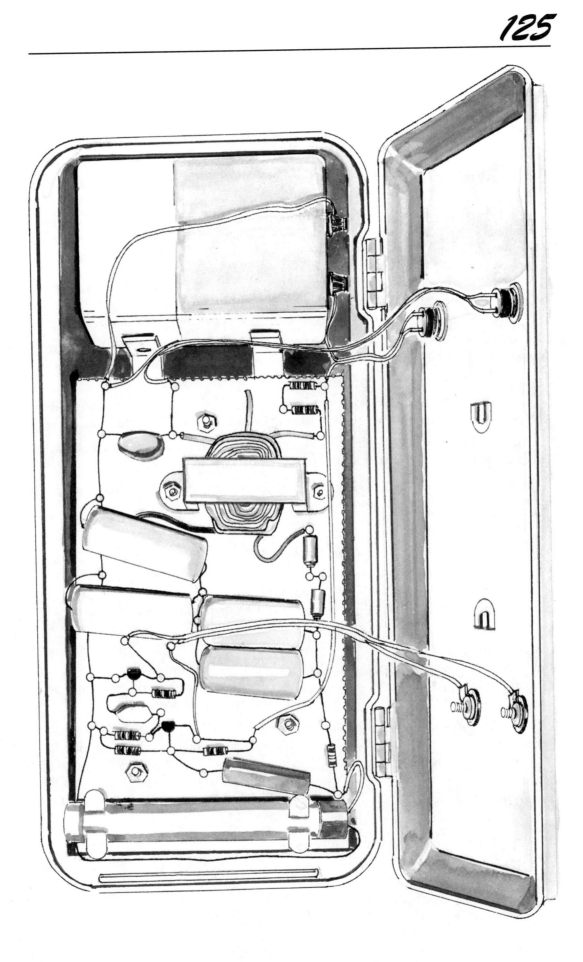

PROCEDURE FOR CHARGING GEIGER COUNTER

1. Needless to say, turn on the power switch (S1).

2. Put on the headphones.

3. Tap the charge switch (S2) firmly in a rapid-fire manner about 60 times, or until you hear random clicking. This is normal background noise caused by cosmic rays.

4. After the unit is charged up, tap the switch occasionally to keep it charged. You'll be able to tell how often this has to be done when you start using the counter. It may range from several seconds to a minute or so.

Mount the completed package on rubber mounts in an enclosure. Any kind of enclosure will do; we just happened to pick a metal tool box. If you also use a metal box, be careful not to let any of the exposed wiring touch the metal walls.

Before you install the assembly, cut a large rectangular hole in the wall next to where the Geiger tube will be. The hole will serve as a radiation window. Incidentally, both switches and the headphone pin jacks must be mounted on top of the enclosure.

Assuming that you have a gamma source, note that when you bring the source near the Geiger counter, or vice versa, the clicking will increase dramatically.

If you were fortunate enough to locate more than one source, the counter will indicate their relative strengths by the clicking intensity.

With the Geiger counter and gamma-ray source, you can now do the last part of Experiment No. 7 (on gamma-ray shielding materials).

Geiger Counter Parts List

R1, R2—1.0-ohm, ½-watt carbon resistor

R3—1,500,000-ohm, ½-watt carbon resistor

R4, R5, R7—560,000-ohm, ½-watt carbon resistor

R6—2700-ohm, ½-watt carbon resistor

C1—0.0033-mfd, 600-volt tubular capacitor

C2, C3, C4, C5—0.47-mfd, 600-volt tubular capacitor

C6—0.006-mfd, 1600-volt tubular capacitor

C7—0.047-mfd, 250-volt (or higher) tubular capacitor

T—6.3-volt AC, 1.0 ampere filament transformer

S1—SPST toggle switch

S2—SPST normally open push-button switch

D1, D2—Silicon rectifier, 1000-volt PIV rating, 1-ampere (or higher) rated current capacity, low reverse current (20 microamperes or lower)

Q1, Q2—NPN silicon transistor (2N3402 or equivalent)

Headphones—2000-ohm high sensitivity headphones

B—6-volt battery with screw terminals

G—Geiger tube (A tube and bakelite socket are available for about $20 from The Nucleus Inc., Box R, Oak Ridge, Tennessee 37830. Ask for tube-socket unit SG-2)

Perforated chassis board

Push-in terminals

Hookup wire

Suitable enclosure

HANDING ON THE TORCH

The promises of the many, many other discoveries and inventions made by Edison remained unfulfilled during his lifetime. Some, such as his observation and documentation of the Edison effect and the etheric force, were the underpinnings of entire industries (electronic and the computer) that did not even begin to emerge until a generation after his time.

Thomas Edison and the inventors and scientists of his and other generations that have come and gone—our ancestors—gave us the telephone, the microphone, the transistor and the computer.

We hope that the readers of this book, the potential Edisons of the future, will leave such useful legacies for the next generations of boys and girls everywhere who, in turn, will be our future scientists.

Thomas Alva Edison holding one of his Edison Diamond Disc phonograph records in the library of his West Orange laboratory, September 20, 1916.

**ALPHABETICAL LISTING OF RECOMMENDED
SCIENCE PUBLICATIONS
COMPILED BY
DR. WALTER BISARD, DIRECTOR
SCIENCE & MATH TEACHING CENTER
CENTRAL MICHIGAN UNIVERSITY**

[A]

AMERICAN BLACK SCIENTISTS AND INVENTORS. Edward S. Jenkins and others. NSTA. 1975.

[B]

BASIC GENETICS: A HUMAN APPROACH. Sr.-High & College Level. Student Text (1 copy). Teacher's Edition (1 copy). Kendall/Hunt. 1985.

BATTERIES AND BULBS. Grades 4/6. Teacher's Guide. Delta Eduction, Inc. 1986.

BATTERIES AND BULBS—TEACHER'S GUIDE. Elementary Science Study. McGraw-Hill Book Co., Webster Div. 1975

BEEHIVES OF INVENTION: EDISON AND HIS LABORATORIES. Davidson, G. Office of Publications, National Parks Service. 1973.

THE BEST OF ENERGY BOOK. Grades 1/3 & Grades 4/6. Sampler. McDonald's Corp. 1974.

[C]

A CENTURY OF LIGHT. Cox, J. A Benjamin Co. Inc. 1979.

CHEMICAL DEMONSTRATIONS: A HANDBOOK FOR TEACHERS OF CHEMISTRY (VOL. 1). Sr.-High/College Level. Teacher resource/reference text. The Univ. of Wisconsin Press. 1983.

CHEMICAL DEMONSTRATIONS: A HANDBOOK FOR TEACHERS OF CHEMISTRY (VOL. 2). Sr.-High/College Level. Teacher resource/reference text. The Univ. of Wisconsin Press. 1985.

COME FLY WITH ME. Housel, D.C., & Housel, D.K.M. The Michigan Aerospace Education Council. 1984.

CONCEPTUAL PHYSICS. College Level. Student Text. Little, Brown and Co. 1985.

CONDITIONS FOR GOOD SCIENCE TEACHING. NSTA. 1984.

CONNECTIONS—A CURRICULUM IN APPROPRIATE TECHNOLOGY FOR 5TH AND 6TH GRADES. Melcher. J. National Center for Appropriate Technology. 1980.

CREATIVE SCIENCING: A PRACTICAL APPROACH. (2nd ed.) Devito, A., & Krockover, G.H. Little, Brown and Co. 1980.

[D]

DR. ZED'S ZANY BRILLIANT BOOK OF SCIENCE EXPERIMENTS. Penrose, G. Greey dePencier Publisher. 1977.

[E]

EARLY CHILDHOOD AND SCIENCE. Comp., Margaret McIntyre. NSTA. 1984.

THE EARTH AND BEYOND. Jr./Sr.-High Level. Teacher's Edition. Steck-Vaughn. 1986.

EARTH SCIENCE. Grades 6/9. Teacher's Ed. Scott, Foresman and Co. 1986.

EDISON EXPERIMENTS YOU CAN DO: BASED ON THE ORIGINAL LABORATORY NOTEBOOKS OF THOMAS ALVA EDISON. Harper & Brothers, Publishers. 1960. [2]

EDISON—THE MAN WHO TURNED DARKNESS INTO LIGHT. James G. Cook. Thomas Alva Edison Foundation. Inc. 1986.

ELECTRIC POWER. Grades 5/6. Student ed. Silver Burdett. 1981.

ELECTRICITY. Grades 3/10. Teacher Resource Book. TOPS Learning Systems. 1983.

ELECTRICITY: TODAY'S TECHNOLOGIES, TOMORROW'S ALTERNATIVES. Electric Power Research Institute. 1981.

ELEMENTARY SCHOOL SCIENCE AND HOW TO TEACH IT. (4th ed.) Blough, G.O., & Schwartz, J. Holt, Rinehart and Winston, Inc. 1969.

ENERGY: A PHYSICAL SCIENCE. Grades 7/9. Student ed. (1). Teacher's Manual (1) w/ tests. Harcourt, Brace & Janovitch. 1980.

EXPERIENCES IN PHYSICAL SCIENCE. Middle school level. Teacher's ed. Laidlaw Brothers. 1985.

EXPERIENCES IN PHYSICAL SCIENCE RESOURCES. Middle school level. Teacher's Resources Edition: Activity masters; Test masters; Transparency masters; Study Guide masters; Laidlaw Brothers. 1986.

EXPLORATIONS IN CHEMISTRY. Gray. C. E.P. Dutton & Co., Inc. 1965.

[F]

FOCUS ON EXCELLENCE. NSTA Series (selected texts). Science/Technology/Society. Ed., John E. Penick & Richard Meinhard-Pellens. 1984. Physics. Ed., John E. Penick. 1985. Middle School/Junior-High Science. Ed., John E. Penick & Joseph Krajcik. 1985. Chemistry. Ed., John E. Penick. 1985. Exemplary Programs in Physics, Chemistry, Biology and Earth Science. Ed., Robert E. Yager.

FOSSIL FUELS. Grades 4/12. Student Activities Edition. Research Foundation of the State Univ. of New York, on behalf of the New York Energy Education Project. 1985.

4-H MATERIALS (Cooperative Extension Service of the U.S. Department of Agriculture): ELECTRICITY'S SILENT PARTNER—MAGNETISM: FUNDAMENTALS OF ELECTRICITY—PART II. EXPLORING THE WORLD OF ELECTRICITY: FUNDAMENTALS OF ELECTRICITY—PART 1. INTRODUCTION TO THE WORLD OF ELECTRONICS: UNIT 6.

FROM ALCHEMY TO ACCELERATORS. Chemistry reprints from The Science Teacher. NSTA. 1984.

FROM ATOMS TO INFINITY: READINGS IN MODERN SCIENCE. Ed., Simak. C. Harper & Row, Publishers. 1965.

FUN WITH SCIENCE: EASY EXPERIMENTS FOR YOUNG PEOPLE. Freeman, Mae & Ira. Random House. 1956.

[G]

[H]

[I]

IDEAS AND INVESTIGATIONS IN SCIENCE—EARTH SCIENCE. Berstein, L., & Wong, H.K. Globe Book Company. 1979.

INTRODUCTORY CHEMISTRY: MODELS & BASIC CONCEPTS. Amend, J. John Wiley & Sons. 1977.

[J]

[K]

KINETIC MODEL. Grades 5-12th Level. Teacher's Manual. Student Task Cards. TOPS Learning System. 1978.

[L]

LIGHT. Grades 5-12th Level. Teacher's Manual. Student Task Cards. TOPS Learning Systems. 1978.

[M]

MANAGEMENT OF THE ELECTRIC ENERGY BUSINESS. Vennard, E. McGraw-Hill Book Company. 1979.

METRIC MEASURING. Grades 3-10th Level. Teacher Resource Book. TOPS Learning Systems. 1984.

MODELS: ELECTRIC AND MAGNETIC INTERACTION. Science Curriculum Improvement Study. Delta Education. 1971.

MR. WIZARD'S EXPERIMENTS FOR YOUNG SCIENTISTS. Don Herbert. Jr./Sr.-High Level. Resource Book. Doubleday & Company, Inc. 1959.

MR. WIZARD'S SUPERMARKET SCIENCE. Don Herbert. Jr./Sr.-High Level. Resource Book. Random House. 1980.

[N]

NASA EDUCATIONAL PUBLICATIONS

NATIONAL SCIENCE TEACHERS ASSOCIATION |NSTA| YEARBOOKS. 1983–Science Teaching: A Profession Speaks. Ed., David P. Butts and Faith K. Brown. 1984–Redesigning Science and Technology Education. Ed., Rodger Bybee, Janet Carlson, and Alan McCormack. 1985–Science. Technology, and Society. Ed., Roger Bybee.

NATURE WITH CHILDREN OF ALL AGES. Edith Sisson. Prentice-Hall, Inc. 1982.

NUCLEAR ENERGY. Grades 6-12Th Level. Student Activities. Research Foundation of the State Univ. of New York, on behalf of the New York Energy Education Project. 1985.

[O]

[P]

PENDULUMS Grades 3-10th Level. Teacher Resource Book. TOPS Learning Systems. 1983.

PHYSICAL SCIENCE. Grades 7-9th Level Package: Student Texts (2); Teacher's Annotated Edition (2); Teacher's Annotated Lab Book (1); Teacher's Resource Book (1); Teacher's Resource Center (1). Macmillan. 1986.

PHYSICAL SCIENCE. Grades 6-9th Level. Teacher's Edition. Scott, Foresman and Company. 1986.

PHYSICAL SCIENCE. Jr.-High Level. Student Edition (sampler). Silver Burdett. 1987.

PHYSICAL SCIENCE: AN INQUIRY APPROACH. College Level. Student Edition. Instructor's Manual. Canfield Press. 1977.

PLAYING WITH ENERGY. These classroom games and simulations are for grades 9-12. NSTA. 1981.

POLYMER CHEMISTRY (plus tutorial disk for Apple II). NSTA. 1986.

PROJECT AIMS. Grades K-12. ACTIVITIES FOR INTEGRATING MATHEMATICS & SCIENCE. Fresno Pacific College. 1984.

PROMOTING SCIENCE AMONG ELEMENTARY-SCHOOL PRINCIPALS. Kenneth R. Mechling and Donna L. Oliver. Four Handbooks: Science Teaches Basic Skills-I; The Principal's Role in Elementary School Science-II; Characteristics of a Good Elementary Science; Program-III, What Research Says About Elementary-School Science-IV. NSTA. 1987. 1983.

PSNS/AN APPROACH TO PHYSICAL SCIENCE. College Level. Student Text. Rensselaer Polytechnic Institute. 1969.

[Q]

[R]

RESEARCH ADVENTURES FOR YOUNG SCIENTISTS. Barr, G. McGraw-Hill Book Co. 1964.

RESEARCH WITHIN REACH: SCIENCE EDUCATION. Ed., David Holdzkom and Pamela B. Lutz. NSTA. 1984.

[S]

SAFETY IN THE ELEMENTARY SCIENCE CLASSROOM. NSTA. 1978.

SAFETY IN THE SECONDARY SCHOOL CLASSROOM AND LABORATORY. NSTA. 1987.

SCIENCE. 1st-6th Grades. Student Edition (one per grade level). Teacher's edition (one per grade level). Scott, Foresman and Co. 1986.

SCIENCE...A PROCESS APPROACH II. American Assoc. for the Advancement of Science. Xerox-Ginn and Co. 1974.

SCIENCE AND CHILDREN INDEX DISK. SEPT. 1982–MAY 1986. NSTA. 1986.

SCIENCE EXPERIENCES WITH INEXPENSIVE EQUIPMENT. Jr./Sr.-High Level. Teacher/Student Text. International Textbook Co. 1950.

SCIENCE EXPERIENCES WITH TEN-CENT STORE EQUIPMENT. Jr./Sr.-High Level. Student Edition. International Textbook Co. 1950.

SCIENCE FAIRS AND PROJECTS. NSTA. 1985.

SCIENCE FOR THE ELEMENTARY SCHOOL. Victor, E. The Macmillan Company. 1970.

SCIENCE IN ELEMENTARY EDUCATION. Gega, P.C. Elem. Education. Teacher Reference. John Wiley & Sons, Inc. 1970.

SCIENCE IN EVERYDAY LIFE. Vergara, W.C. Harper and Row. 1980.

SCIENCE ON A SHOESTRING K-7th Level. Teacher Resource Book. Addison-Wesley. 1985.

SCIENCE TRICKS & EXPERIMENTS. K-12th Level. Commerical Publication. Radio Shack. 1984.

THE SCIENTIFIC AMERICAN BOOK OF PROJECTS FOR THE AMATEUR SCIENTIST. Stong, C.L. Simon and Schuster. 1960.

THE SEARCH FOR SOLUTIONS. Judson, H.F. Holt, Rinehart and Winston. 1980.

THE SEARCH FOR SOLUTIONS: TEACHING NOTES. Jr./Sr.-High Level. Teacher's Guide. Reeves Corporate Services for Phillips Petroleum Public Affairs. 1980.

SILVER BURDETT BIOLOGY. Sr.-High Level. Teacher's Resource Package: Teacher's Edition; Student's Edition; Overhead Copy Masters; Copy Masters. Silver Burdett Co. 1986.

SMALL THINGS: AN INTRODUCTION TO THE MICROSCOPIC WORLD. Elementary Level. Teacher's Guide. McGraw-Hill, Webster Div. 1967.

A SOURCEBOOK FOR THE PHYSICAL SCIENCES. Joseph et al. Harcourt, Brace & World, Inc. 1961.

[T]

TEACHING CHILDREN ABOUT SCIENCE. Elaine Levenson. Prentice-Hall, Inc.

TEACHING THE FUN OF PHYSICS. Janice Pratt Van Cleave. Elementary Level. Idea Catalog for Teachers. Prentice-Hall, Inc. 1985.

TEACHING SCIENCE AS CONTINUOUS INQUIRY: A BASIC 2-E. Rowe, M.B. McGraw-Hill. 1978.

TEACHING SCIENCE WITH EVERYDAY THINGS. (2nd ed.) Schmidt, V., & Rockcastle, V. McGraw-Hill. 1982.

THOMAS EDISON, CHEMIST. Vanderbilt, B. American Chemical Society. 1971.

THOMAS EDISON: THEY CALLED HIM WIZARD. Cunningham, J. Public Service Electric and Gas Company, Newark, NJ. 1979.

3-2-1 CONTACT TEACHER'S GUIDE. Elementary/Jr.-High Level. Teacher's Guide. Children's Television Workshop. 1985.

[U]

NEW UNESCO SOURCEBOOK FOR SCIENCE TEACHING. UNESCO. 1974.

[V]

[W]

WHAT RESEARCH SAYS TO THE SCIENCE TEACHER. VOL. 1-1978, ed., Mary Budd Rowe; VOL. 2-1979, ed., Mary Budd Rowe; VOL. 3-1981, ed., Norris C. Harms & Robert E. Yager; VOL. 4-1982, ed., Robert E. Yager. NSTA.

[X, Y, Z]

VIDEOTAPES, CASSETTES, FILMS, DISKS, and SLIDES

SCIENCE TOOLKIT; MASTER MODULE. Jr. High—College Level Resource Guide Package: User's Manual & Experiment Guide; Interface Box; Thermistor; Photocell. Broderbund Software, Inc. 1985.

SEARCH FOR SOLUTIONS: ADAPTION, CONTEXT, TRIAL AND ERROR. (VHS)

SEARCH FOR SOLUTIONS: INVESTIGATION, EVIDENCE, PATTERNS. (VHS)

SEARCH FOR SOLUTIONS: MODELING, THEORY, PREDICTION. Jr. High-College Level. VHS Tape. Phillips Petroleum Co. 1980.

YOU CAN BE A SCIENTIST TOO! (2) level VHS videotape (12:51). The Equity Institute. 1985.

Resting on the doorstep of Mr. Edison's laboratory at Fort Myers, March 15, 1931: Henry Ford, Thomas A. Edison, and Harvey S. Firestone.

CHRONOLOGY OF EVENTS
IN THE LIFE OF
THOMAS ALVA EDISON

1847 February 11—born in Milan, Ohio, son of Samuel and Nancy Elliott Edison.

1854 Edison family moved to Port Huron, Michigan.

1859 A newsboy and "candy butcher" on the train of the Grand Trunk Railway running between Port Huron and Detroit.

1862 Printed and published a newspaper, *The Weekly Herald*, on the train—the first newspaper ever printed on a moving train.

1862 August—saved from death the young son of J. U. MacKenzie, station agent at Mt. Clemens, Michigan. In gratitude, the father taught Edison telegraphy.

1863 May—first position as a regular telegraph operator on Grand Trunk Railway at Stratford Junction, Ontario, Canada.

1863 Began a five-year period during which he served as a telegraph operator in various cities of the central western states, always studying and experimenting to improve apparatus.

1868 Made his first patented invention—the electrical vote recorder. Application for patent signed October 13, 1868.

1869 Landed in New York City, poor and in debt. Shortly afterwards, looking for work, was in operating room of the Gold Indicator Company when its apparatus broke down. No one but Edison could fix it, and he was given a job as superintendent.

1869 October—established a partnership with Franklin L. Pope as electrical engineers.

1870 Received his first money for an invention—$40,000 paid him by the Gold and Stock Telegraph Company for his stock ticker. Opened a manufacturing shop in Newark where he made stock tickers and telegraph instruments.

1871 December 25—married Mary Stilwell, daughter of Nicholas Stilwell, of Newark, New Jersey.

1872 Began a four-year period during which he conducted manufacturing of telegraph instruments for Western Union Telegraph Company and Automatic Telegraph Company. He had several shops during this time in Newark, New Jersey. He worked on and completed many inventions, including the motograph, automatic telegraph system, duplex, quadruplex, and multiplex telegraph systems; also paraffin paper and the carbon rheostat.

1874 Invented the "Electric Pen" and manual duplicating press for making copies of letters.

1875 Devised an automatic copying machine. In Edison's own words, "...toward the latter part of 1875 I invented a device for multiplying copies of letters called the Mimeograph." He later sold the machine to A. B. Dick of Chicago.

1875 November 22—discovered a previously unknown and unique electrical phenomenon, which he called "etheric force." Twelve years later, this phenomenon was recognized as being due to electric waves in free space. This discovery is the foundation of wireless telegraphy.

1876 April—moved from Newark to his newly constructed laboratory at Menlo Park, New Jersey. This was the first laboratory for organized industrial research.

1877 April 27—applied for patent on the carbon telephone transmitter, which made the telephone commercially practicable. This invention included the microphone, which is used in radio broadcasting.

1877 December 6—recorded "Mary Had a Little Lamb" on the tin-foil phonograph. This was the first time a machine had recorded and reproduced sound.

1877 December 24—applied for patent on the phonograph.

1878 April 18—took the tin-foil phonograph to Washington, D.C. to demonstrate it before the National Academy of Sciences and to President Rutherford B. Hayes and White House guests.

1878 July 29—using the heat of the sun's corona during an eclipse at Rawlins, Wyoming, he tested microtasimeter, a device indicating minute heat variations by electrical means.

1878 October 15—incorporation meeting of the Edison Electric Lighting Company.

1879 July—first Edison experimental marine electrical plant installed aboard S. S. *Jeanette* for the George Washington De Long expedition to the Arctic.

1879 Invented the first practical incandescent electric lamp on October 21. The lamp maintained its incandescence for more than 40 hours.

1879 Invented radical improvements in construction of dynamos, making them suitable for generators for his system of distribution of current for light, heat, and power. Invented systems of distribution, regulation, and measurement of electric current, including sockets, switches, fuses, etc.

1879 December 31—gave a public demonstration of his electric lighting system in streets and buildings at Menlo Park, New Jersey.

1880 Discovered a previously unknown phenomenon. He found that an independent wire or plate placed between the legs of the filament of an incandescent lamp acted as a valve to control the flow of current. This became known as the "Edison Effect." This discovery covers the fundamental principle on which rests the science of electronics.

1880 April 3—invented the magnetic ore separator.

1880 May 1—first commercial installation of the Edison lighting system of land or water installed on the *S. S. Columbia.*

1880 May 13—started operation of the first passenger electric railway in this country at Menlo Park, New Jersey.

1880 Ushered in seven strenuous years of invention and endeavor in extending and improving the electric light, heat, and power systems. During these years he took out more than 300 patents. Of 1093 patents issued to Thomas A. Edison, 365 deal with electric lighting and power distribution.

1880 October 1—first commercial manufacture of incandescent lamps began at Edison Lamp Works, Menlo Park, New Jersey.

1881 January 31—opened offices of the Edison Electric Light Company at 65 Fifth Avenue, New York City.

1881 March 2—Edison arranged to open the Edison Machine Works at 104 Goerck Street, New York City.

1882 January 12—opened the first commercial incandescent lighting and power station at Holton Viaduct, London, England.

1882 May 1—moved the first commercial incandescent lamp factory from Menlo Park to Harrison, New Jersey. Organized and established shops for the manufacture of dynamos, underground conductors, sockets, switches, fixtures, meters, etc.

1882 September 4—commenced the operation of the first commercial central station for incandescent lighting in this country at 257 Pearl Street, New York City.

1883 July 4—first three-wire system central station for electric lighting started operation at Sunbury, Pennsylvania.

1883 November 15—filed patent on an electrical indicator using the Edison Effect. This was the first patent in the science now known as electronics.

1884 August 9—his wife, Mary Stilwell Edison, died at Menlo Park, New Jersey.

1885 March 27—patent executed on a system for communicating by means of wireless induction telegraphy between moving trains and railway stations.

1885 May 14—patent executed on a ship-to-shore wireless telegraphy system by induction.

1886 January—bought Glenmont, a residence in Llewellyn Park, West Orange, New Jersey.

1886 February 24—Married Mina Miller, daughter of Lewis Miller, of Akron, Ohio.

1886 December—moved plant of Edison Machine Works from 104 Goerck Street, New York City to Schenectady, New York.

1887 November 24—moved his laboratory to West Orange, New Jersey. During the first four years of his occupancy of the West Orange laboratory, he took out over 40 patents concerning improvements on the cylinder phonograph.

1889 October 6—first projection of an experimental motion picture. This was a "talkie" shown at the West Orange laboratory; the picture was accompanied by synchronized sound from a phonograph record.

1891 August 24—applied for patent on the motion picture camera. With the invention of this mechanism, which used a continuous tape-like film, it became possible to take and reproduce motion pictures as we do today.

1891 This year marked the culmination of his preliminary surveys and experimental work on iron-ore concentration that he had started while in Menlo Park in 1880. Edison did some of his most brilliant engineering work in connection with this project.

1893 Edison-Lalande primary cells supplied power for the first electric semaphore signal installed on railroad near Phillipsburg, New Jersey.

1894 April 14—first commercial showing of motion pictures took place with the opening of a peephole Kinetoscope parlor at 1155 Broadway, New York City.

1896 Experimented with the X-ray discovered by Roentgen in 1895. Developed the fluoroscope, which invention Edison did not patent, choosing to leave it to public domain because of its universal need in medicine and surgery.

1896 February 26—his father, Samuel Edison, died in Norwalk, Ohio.

1896 April 23—first commercial projection of motion pictures at Koster & Bial's Music Hall, New York City, by the Edison Vitascope.

1896 May 16—applied for a patent on the first fluorescent electric lamp. This invention sprang directly from his work on the fluoroscope.

1900 This year marked the beginning of a ten-year period of work that resulted in the invention of the Edison nickel-iron-alkaline storage battery and its commercial introduction. The alkaline battery is widely employed as a power source in mine haulage, inter- and intra-plant transportation, for railway train car lighting and air conditioning, signaling services, and many other industrial applications.

1901 Commenced construction on the Edison cement plant at New Village, New Jersey, and started quarrying operations at nearby Oxford. In his cement industry, Edison proceeded to apply the fruits of experience gained in the iron-ore concentrating venture.

1902 Worked on improving the Edison copper oxide primary battery.

1903 July 20—applied for patent on long rotary kilns for cement production.

1907 Developed the universal electric motor for operating dictating machines on either alternating or direct current.

1910 This year initiated a four-year period of work on an improved type of disc phonograph. His work resulted in production of the "Diamond Disc" instrument and records, which reproduced vocal and instrumental music with improved fidelity.

1912 Introduced the Kinetophone for talking motion pictures.

1914 October 13—patent executed on electric safety lanterns, which are used by miners for working lights. These miners' lamps have contributed in an important degree to the reduction of mine fatalities.

1914 Developed a process for the manufacture of synthetic carbolic acid. Designed a plant, and within a month was producing a ton a day to help overcome the acute shortage due to the World War.

1914 December 9—Edison's great plant at West Orange, New Jersey, was destroyed by fire. Immediate plans for rebuilding were laid, and new buildings began to arise almost before the ruins of the old ones were cold.

1914 Invented the Telescribe, combining the telephone and the dictating phonograph, thus permitting the recording of both ends of telephone messages.

1915 Established plants for the manufacture of fundamental coal-tar derivatives vital to many industries previously dependent on foreign sources. These coal-tar products were needed later for the production of wartime explosives. Edison's work in this field is recognized as having paved the way for the important development of the coal-tar chemical industry in the United States today.

1915 October 7—became President of the Naval Consulting Board, at the request of Josephus Daniels, then Secretary of the Navy. During the war years (1915-1918), he did a large amount of work connected with national defense, particularly with reference to special experiments on over 40 major war problems for the United States Government.

1923 Made a study of economic conditions, the result of which was published in a pamphlet in 1924, when Edison presented to the Secretary of the Treasury a proposed amendment to the Federal Reserve Banking System.

1927 A four-year period began during which Edison searched for a domestic source of natural rubber. This project was completed with the vulcanization of goldenrod rubber shortly before his death.

1928 October 20—received a special Congressional Medal, which was presented by Andrew W. Mellon, Secretary of the Treasury.

1929 October 21—commemorating the 50th anniversary of the incandescent lamp and in the presence of President Herbert Hoover, Henry Ford, and other world leaders, Edison re-enacted the making of the first practical incandescent lamp.

1931 October 18—died at Llewellyn Park, West Orange, New Jersey at the age of 84. He was survived by his wife, Mina Miller Edison; his four sons, Thomas Alva, Jr., William Leslie, Charles, and Theodore; and his daughters Marion Edison Oser and Madeleine Edison Sloane.

INDEX

A

Alarm, burglar alarm, making of, 45
Atoms
   decay process, 112-113
   explanation of, 112
   isotopes, 113
   radium atom, 12
   splitting, oil-drop model, 113-114

B

Battery
   electric, 7
   making of, 61-63
Bell, Alexander Graham, 17
Burglar alarm, making of, 45

C

Camera, pinhole camera, 41-42
Carbon experiments
   carbon dioxide, 11-13
   carbon, variable conductivity of, 16
   carbon transmitter, 16-18
      building of, 18
      principle, 16
      testing of, 18
Chain reaction, domino model,
   114-115
Chemical experiments
   battery, making of, 61-63

   carbon dioxide, 11-13
   crystals, from sugar solution, 13
   ink, invisible, 10-11
Cigar-box microphone, 67-69
Circuits
   doorbell, 2
   electrical, simple, 1
   series and parallel, 23-24
Coal
   conversion to fuel gas, 105-106
   methane, making of, 104-105
Code set, making of, 56-59
Coins, detecting counterfeit coins,
   39-40
Commercial laboratory, development
   of, 69
Compass, and magnetism, 21-22
Conductors/insulators, 3-4
Crystals, from sugar solution, 13
Current, controlling with pencil, 4-5

D

Decay process, atoms, 112-113
Direct-current generator, 46-49
Drawer lock, making of, 37-39

E

Edison, Thomas
   characteristics of, 33
      learning about past discoveries, 49

Edison, Thomas (*con't.*)
  observational skills, 15
  persistence, 49
  chemistry
    commercial laboratory,
      development of, 69
    love of, 59
  Faraday, Michael, influence of,
    25-26, 49
  improving inventions of others
    telegraph, 19-20, 34
    telephone, 17, 66-67
Electrical experiments
  circuits
    doorbell, 2-3
    series and parallel, 23-24
    simple electrical, 1-2
  conductors/insulators, 3-4
  current, controlling with pencil, 4-5
  electrolyte, 5-6
  electrophorus, building of, 20
  electroscope, building of, 21
  fuse, 24-25
  ice as conductor of electricity,
    29-30
  "ice pail" experiment, 27-29
  two-way switch, 2-3
  See also Carbon experiments.
Electricity
  converting from wind energy, 98-99
  transforming from sunlight, 95-96
Electric light
  direct-current generator, 46-49
  light-bulb indicator, 43-44
  making, 43
Electric pen, 51-54
  making of, 53-54
Electrochemical experiments
  electric battery, 7-9
  electricity, from a lemon, 6-7
  electroplating, housekey, 30-31
  gases, from electrified saltwater,
    9-10
Electrolysis, 10
Electrolyte, 5-6
Electromagnetic experiments
  coins, detecting counterfeit coins,
    39-40
  drawer lock, 37-39
  electromagnet
    making, 21
    with two coils, 22-23
  ore separator, 39
  telegraph relay, making of, 35-37
Electrophorus, building of, 20
Electroscope, building of, 21

Energy
  conversion as lost energy, 72
  forms of, 71
  laws of thermodynamics, 71
Energy (alternate forms) experiments
  coal
    conversion to fuel gas, 105-106
    methane, making of, 104-105
  fuel cell, making of, 108-109
  geothermal energy experiment,
    106-108
  nuclear experiments, 111-127
  ocean-thermal energy experiments,
    100-101
  solar energy experiments, 86-96
  trash, converting to energy,
    102-103
  wind energy experiments, 98-99
  See also specific categories of
    experiments.
Energy conservation experiments
  draft detector, making of, 75
  energy-saving experiment, 78-81
  forces of nature and home, 73-74
  home energy audit, 74-77
  weatherizing model house, 81-85
  See also Solar energy.

F

Fuel
  converting coal to, 105-106
  fuel cell, making of, 108-109
Fuel cell, compared to battery, 108
Fuse, 24-25

G

Garden, solar garden, building of,
  86-91
Gases, from electrified saltwater, 9-10
Geothermal energy experiment, steam
  engine, making of, 106-108

H

Home energy audit, 74-77
Hot-water heater, 94-95

I

Ice as conductor of electricity, 29-30

"Ice pail" experiment, 27-29
   electrophorus for, 20
   electroscope for, 21
Ink, invisible, 10-11
Insulation, weatherizing model house, 81-85
Insulators/conductors, 5
Invisible ink, 10-11
Isotopes, 113

K

Kinetoscope, 40

L

Light
   direct-current generator, 46-49
   electric light, making, 43
   light-bulb indicator, 43-44

M

Magma, 105
Magnetism experiments
   compass, and magnetism, 21-22
   electromagnet
      making, 21
      with two coils, 22-23
   magnet, identifying poles, 20-21
Methane, making from coal, 104-105
Microphone, cigar-box microphone, 67-69
Motion-picture related experiments
   persistence-of-vision effect, 41
   pinhole camera, 41-42

N

Nuclear energy
   atoms
      decay process, 112-113
      explanation of, 112
      isotopes, 113
      radium atom, 112
   uranium, 111
Nuclear experiments, 111-127
   chain reaction, domino model, 114-115
   Geigercounter, building of, 125-126
   nuclear power plant steam turbine, 120-121

radioactivity
   observing by radiography, 117-118
   observing with cloud chamber, 118-120
   observing with electroscope, 115-117
   shielding of, 121-122
   splitting atom, oil-drop model, 113-114

O

Ocean-thermal energy experiments, ocean thermal-energy conversion, 100-101
Ore separator, 39

P

Parallel/series circuits, 23-24
Pen, electric pen, 51-54
Persistence-of-vision effect, 41
Phonograph, pickup, making of, 34-35
Pinhole camera, 41-42

R

Radio, making of, 64-66
Radioactivity
   observing by radiography, 117-118
   observing with cloud chamber, 118-120
   observing with electroscope, 115-117
   shielding of, 121-122

S

Solar energy, 93-94
   advantages of, 94
Solar energy experiments
   electricity, transforming from sunlight, 95-96
   hot-water heater, 94-95
   solar garden, building, 86-91
   water as heat storage experiment, 91
Sound experiments, phonograph pickup, making of, 34-35
Splitting atom, oil-drop model, 113-114

Steam engine, geothermal model,
    106-108
Sunlight. *See* Solar energy
    experiments.
Switch, two-way, 2-3

T

Telegraph
    code set, making of, 56-59
    Edison's improvement of, 19-20, 34
    relay, making of, 35-37
    See also Magnetism experiments.
Telephones
    Edison's improvement of, 17, 66-67
    See also Carbon experiments.

Transmitter, carbon transmitter, 16-18
Trash, converting into usable energy,
    102-103
Two-way switch, 2-3

W

Water as heat storage experiment, 91
Weatherizing model house, 81-85
Wind energy, 97-98
    problems related to, 97-98
Wind energy experiments, electricity,
    converting from wind energy, 98-99